Electronic Banking

Edited by SCN Education B.V.

HOTT Guide

Edited by SCN Education B.V.
This series of books cover special topics which are useful for a business audience. People who work or intend to work with Internet - at a management, marketing, sales, system integrating, technical or executive level - will benefit from the information provided in the series.

These books impart how new technologies and sales & marketing trends on Internet may be profitable for business. The practical knowhow presented in this series comes from authors (scientists, research firms and industry experts, a.o.) with countless years of experience in the Internet area.

The HOTT Guides will help you to:
- Enlarge your knowledge of the (im)possibilities of Internet and keep it up-to-date
- Use Internet as an effective Sales & Marketing tool by implementing new technologies as well as future-oriented strategies to improve your business results
- Facilitate decision making on a management level
- Reduce research costs and training time

These books are practical 'expert-to-manager' guides. Readers will see a quick 'return on investment'.

The Publishing department of SCN Education B.V. was founded in 1998 and has built a solid reputation with the production of the HOTT Guide series. Being part of an international IT-training corporation, the editors have easy access to the latest information on IT-developments and are kept well-informed by their colleagues. In their research

HOTT Guide

activities for the HOTT Guide series they have established a broad network of IT-specialists (leading companies, researchers, etc) who have contributed to these books.

SCN Education B.V., Newtonstraat 37C, 3902 HP Veenendaal, The Netherlands,
tel. +31 – 318 – 547000, fax +31 – 318 – 549000, www.scnedu.com,
www.hottguide.com.

Books already in print:

Webvertising
Mobile Networking with WAP
ASP – Application Service Providing
Customer Relationship Management
Data Warehousing

Electronic Banking

The Ultimate Guide to Business and Technology of Online Banking

Edited by SCN Education B.V.

vieweg

Die Deutsche Bibliothek - CIP-Cataloguing-in-Publication-Data
A catalogue record for this publication is available from Die Deutsche Bibliothek
(http://www.ddb.de).

Trademarks
All products and service marks mentioned herein are trademarks of the respective owners mentioned in the articles and or on the website. The publishers cannot attest to the accuracy of the information provided. Use of a term in this book and/or website should not be regarded as affecting the validity of any trademark or service mark.

1st Edition April 2001

Vieweg is a company in the specialist publishing group BertelsmannSpringer.

Printed on acid-free paper

ISBN-13: 978-3-322-86629-5 e-ISBN-13: 978-3-322-86627-1
DOI: 10.1007/978-3-322-86627-1

Preface

Net Banking Unveiled

The world of banking and financial services is in the midst of dramatic change. Driven by deregulation, consolidation and information technology, the financial services industry is busy reshaping itself — moving away from traditional "brick-and-mortar" branches to focus on new delivery channels, improved customer service and 24-hour-a-day access to information and transactions.

According to IBM Consulting Group, within 10 years, Internet will become the dominant delivery channel for financial services, even if currently Internet banking operations represent approximately 5% of transactions total volume in the USA and in Europe.

As the industry changes, financial institutions are experimenting with and embracing a wealth of new technologies. Through automated teller machines, Internet banking, extended-hour call centers and more, banks and other operations are educating their consumers on how advanced technology can lower costs while improving the level of customer service. "Bankers' hours" used to be from 9 a.m. to 4 p.m., Monday through Friday. With the help of new technology, today's consumers can check their account balances, apply for loans, get cash or conduct transactions 24 hours a day, seven days a week.

This HOTT Guide reveals all the ins and outs of this new online phenomenon. You are invited to read the expert opinions from bankers, trendwatchers and financial consultants, and explore the new banking solutions through white papers and reports.

The Editors

Acknowledgements

Many people and professionals have contributed directly or indirectly to this book. To name them all would be practically impossible, as there are many. Nevertheless the editors would like to mention a few of those who have made the production of this book possible.

Executive Editor for SCN Education B.V.: *Robert Pieter Schotema*
Publishing Manager: *drs. Marieke Kok*
Product Manager: *Martijn Robert Broersma*
Editorial Support: *Dennis Gaasbeek*
HOTT Guide Online: *Rob Guijt, Richard van Winssen*
Interior Design: *Paulien van Hemmen, Bach.*

Contents

Contents

Chapter 1: Introduction to Electronic Banking

The Evolution of Online Investment Banking

Title: *The Evolution of Online Investment Banking*

Author: *Troy J. Strader, Management Information Systems, Iowa State University, and Richard B. Carter, Department of Finance, Iowa State University*

Abstract: *When companies are considering an IPO they must first evaluate the financial issues to decide whether it is a viable financing option, and second they must identify which channel(s) they wish to use to distribute the IPO. In this article we focus on the second decision.*

Biography: Troy J. Strader is Assistant Professor of Management Information Systems at Iowa State University. He received his Ph.D. in Business Administration (Information Systems) from the University of Illinois at Urbana-Champaign in 1997. His research interests include electronic commerce, strategic impacts of information technology, and information economics.
Richard B. Carter is Professor of Finance at Iowa State University. He received his Ph.D. in Finance from the University of Utah in 1987. His research interests include capital acquisition, agency costs and small business.
Email: E-mail: Troy Strader
Web site: Iowa State University, College of Business

Introduction

In the past few years there has been a growth in Internet markets where companies and investors can buy and sell initial public offerings (IPOs) of corporate stock. In fact, by early 1999, the mechanics of Internet IPOs had quickly progressed from the first phase, a partial distribution of IPO shares directly to investors via the Internet, to the actual determination of the offer price and the allocation of shares through an online auction process.

Internet-based investment banks provide companies with a choice of whether to use a traditional investment bank, or one that provides the new online services to distribute some portion of their IPO. When companies are considering an IPO they must first evaluate the financial issues to decide whether it is a viable financing option, and second they must identify which channel(s) they wish to use to distribute the IPO. In this article we focus on the second decision. This is an important topic because it potentially affects all public companies, or companies considering going public, the investment banking industry, and all stock investors. The importance is also indicated by the fact that in 1999 a record $74 billion was raised through 511 initial public stock offerings with new issues posting an average first-day gain of 68.3 percent [Kapadia 2000].

Traditional IPO Process

The traditional IPO process involves the issuing firm, an investment bank that acts as an intermediary between the seller and buyers, and a select group of typically larger investors. The investment bank provides services such as pricing the stock, forming syndicates of investment banks and their brokerage arms to distribute shares, providing access to a select group of large investors to facilitate distribution and, if need be, supporting prices in the IPO after-market by placing its own buy orders for the stock. Prior to the offer, the investment bank contacts its buying clientele and explains the details of the offer and the selling company. During this time the investment bank assesses interest in the IPO and takes preliminary subscriptions for shares. The bank then uses this information to determine the price and the number of shares to sell. Because many IPOs are over-subscribed, the bank pro-rates the shares during the final distribution based on the original subscriptions. This service comes at a price, however, as the investment bank receives a commission -- typically based on the amount of money raised in the IPO.

This process has been used for IPOs for well over a century, but some questionable activities have evolved during that time. There is the practice of spinning, where the investment bank allocates shares to favored or potential customers in hopes of winning future business. One could argue that by spinning, investment banks preclude the average investor from some potentially attractive IPOs. There is also underpricing. The stock price run-up of the average IPO on the first day of trading is so great, that it appears that investment banks are often setting the offer price too low. Theories have emerged to explain the existence and magnitude of underpricing and defend it as an efficient way to clear the IPO market. However, there is still a real possibility that many companies are being sold too cheap.

Consider the case of Theglobe.com, a Website builder that debuted in February 1999. Theglobe's bankers, Bear Stearns and Volpe Brown Whelan, underwrote its shares for $9, raising $27.9 million in capital. On the first day of trading, the price rose to $63.50. Had Theglobe sold the IPO for $63.50, rather than $9, the company would have collected not $27.9 million but $197 million - seven times the money to build the brand and develop new products [Tully 1999]. Given these transaction costs and a less than open IPO market, a new information technology enabled IPO may offer a solution.

Online Investment Banks and the Changing Role of Intermediaries

The primary difference between the traditional and online IPO process is the role of the intermediary. Bakos (1998) identified eight functions of a market that are facilitated through intermediaries. They include determination of product offerings, search, price discovery, logistics, settlement, trust, legal, and regulatory. The differences in the anticipated phases in the diffusion of online IPO processes can be described using these intermediary roles. The phases occur as the participants in the process search for greater market efficiency (reduced transaction costs).

The roles of intermediaries in Phase I, partial share distribution via the Internet, include providing market access and IPO information (search), IPO share offer price information (price discovery), as well as informing the investor that the intermediary feels that the selling company is reputable (trust). For Phase II, price determination via the Internet, the search and trust roles are similar, but the share price is determined not by an offer price but through an auction or negotiation mechanism. The future may bring a Phase III, disintermediation, where the IPO process intermediary is no longer necessary. This phase will only take place when the trust between the share buyer and seller can be facilitated without an intermediary firm.

Each form of online IPO involves the same seller, but a different form of intermediary. The online investment bank provides an Internet-based IPO offering a more open IPO market with access to a larger number of smaller investors. Bob Lessin, CEO of Wit Capital, identified this as a primary goal: to level the Wall Street playing field by giving the little guy, individual investors, a chance to invest in a company when it first offers shares to the public and before the stock actually begins trading in the markets [Dorsey 1998]. An example of a phase I online IPO process is Wit Capital. It allows the investor to subscribe to shares at the offer price via the Internet, using Wit Capital's homepage to peruse pertinent documents concerning the issuing firm. While only a small portion of shares is now allocated to those online investment banks in the distributing syndicate, it appears to be expanding.

The next phase of the online IPO process has been developed by William Hambrecht, owner of W. R. Hambrecht & Co. Using Mr. Hambrecht's plan, dubbed OpenIPO, investors submit bids for the number of shares they would take and at what price. After a few weeks of taking bids, the offering price is set at the lowest price at which all shares can be sold. Those bidding above the offering price will get all the shares they asked for at the offering price; those bidding at the offering price will get a portion of their bid; and those bidding less than the offer price will not get any shares. No more than ten per cent of the shares sold can go to a single bidder, and Hambrecht reserves the right to limit the purchase of anyone seeking to buy more than one per cent [Bransten and Wingfield 1999].

As of the end of 1999 there has not been a significant amount of IPO share sales directly from companies to investors without any intermediary. But, the potential savings in fees and commissions for both the offering company and the investors may be large enough to move toward a disintermeidated process.

Summary and Conclusions

Raising equity in public markets involves many choices for the issuing firms. Among these choices is how much stock to offer and at what price. They must also decide whether to use an investment bank to underwrite the issue and, if so, which investment bank. Traditionally, the underwriter pre-sells the entire offer to its clients - thus determining an optimal offer price and the demand for the issue. However, this traditional method has led to some questions as to some of the practices of the underwriter. For example, is the offer price discounted in an effort to satisfy the underwriter's preferred customers? While there are theories describing deep discounting as an equilibrium market clearing process, the practice suggests that the firm may not have received an optimal price for its stock.

Recently, the IPO process has experienced some significant changes involving the Internet In the first phase, a portion of the issue was distributed via the Internet. A preliminary analysis suggests there has been no fundamental changes in the underlying mechanics of the IPO. In the second phase, both the offer price and the investors are determined through an online auction mechanism. This may result in a more efficient offer price for a company's IPO and certainly implies a gradual disintermediation process as the need for brokerage and other solicitation activities of investment banks is mitigated. However, if deep discounting is required to clear IPO markets, an analysis may uncover the same phenomenon regardless of the price discovery and distribution mechanism.

In the future, when firms are deciding how to distribute their shares, they must compare the risks associated with using a new, relatively untested, online investment bank to distribute a portion of an IPO versus the potential increase in capital received resulting from this new process.

Guide to Online Banking

Title: *Guide to Online Banking*
Author: *from the editors at Womensfinance.com*
Abstract: *A guide to online banking for female consumers. An article on the advantages, disadvantages, methods, services related to online banking.*

Biography: WomensFinance.com educates and energizes women by providing comprehensive solutions and guidance for women's everyday financial needs. Women will discover the freedom and power that comes from making informed financial decisions and putting their own affairs in order regardless of their age or stage in life. Finally, a "one-stop-shop" for all the information women need to make financial decisions online at http://www.womensfinance.com.

Banking & Savings

Guide to Online Banking

With the development of online banking, you no longer have to do all your banking in person. No more waiting in lines. No dealing with the silly vacuum tubes again.

Online banking lets you sit at home in your pajamas and check your accounts 24 hours a day, 7 days a week. You can transfer money from one online account to another or make sure that a certain check has cleared. In a way, your personal computer becomes a virtual teller, ready to assist you with your banking needs.

You may hear online banking referred to as Electronic Banking, Home Banking, Internet Banking, or PC Banking. They are simply different names for the same product.

In order to bank online, you'll need to have two things: a personal computer and a modem connected to a phone line. If you use your bank's software, you'll normally have to use a dial-up connection to access your accounts directly from the bank's computer. Otherwise you'll have to go to your bank's web site on the Internet. You'll need to have an Internet connection and a web browser (Microsoft Internet Explorer or Netscape Navigator) installed on your computer.

Online Banking Methods

There are three ways for you to bank online:

1) Internet-Based.
You access account information through your bank's web site via modem using a standard web browser such as Microsoft Internet Explorer or Netscape Navigator.

2) Bank Software.
Your bank will provide you with proprietary software to install on your computer. The software allows you to connect to your bank's computers over a secure network. You're essentially dialing directly into the bank instead of through the Internet.

3) Personal Finance Software. Software such as Quicken or Microsoft Money allow you to exchange financial information with your bank. The software connects to the bank's computers via an Internet-based setup and downloads financial information from your account. Many banks affiliate leading software vendors such as Intuit or Microsoft to make their systems compatible.

Online Banking Services

When you set up online banking, you can expect some of the following services to be included with your plan. Some are basic. Others features are advanced and may require higher fees to set up. Check with your bank to see what services are offered and if any fees apply.

Things you can do:

- Check account balances.
- Make transfers between online accounts.
- Check the status of loans.
- Track investments.
- Apply for loans online.
- Pay bills electronically.
- Trade stocks or other investments.
- Download account information to a personal finance software program.

Advantages and Drawbacks

Online banking offers many advantages over traditional banking methods. There are also many drawbacks.

1) Advantages

- You can bank from any location that has a Internet connection (Internet-based).
- You can work offline (bank program or personal finance software) after you download account information.
- You have better control over your money.
- You can see which checks have cleared.
- You have access to your account 24 hours a day, 7 days a week.
- You can monitor your account in real time. Withdrawals, deposits, ATMs, debit card purchases.
- You can transfers funds from one account to another.
- You can use financial software to keep track of your online bank account.
- You can set up electronic bill payments.
- Online banking is more cost effective for banks. This should result in lower fees for customers.

2) Drawbacks

- You have to own a personal computer with a modem.
- You must have an internet connection to access Internet-based accounts.
- Your ability to access your bank account is dependent on technology.
- Online banking is subject to the reliability of web servers and other computers. If they crash you can't access your cash.
- You have to know how to use a computer.
- If you ever change banks, you may have to switch software. You'll have to re-enter all your old account information into a new program. This could be very time consuming.

Finding the right bank

Title: *Finding the right bank*
Author: *from the editors at Womensfinance.com*
Abstract: *A list of aspects that have to be taken into account by the consumer in order to choose the right bank.*

Biography: WomensFinance.com educates and energizes women by providing comprehensive solutions and guidance for women's everyday financial needs. Women will discover the freedom and power that comes from making informed financial decisions and putting their own affairs in order regardless of their age or stage in life. Finally, a "one-stop-shop" for all the information women need to make financial decisions online at http://www.womensfinance.com.

Banking & Savings

Finding the Right Bank

If you're like most people, you probably picked your current bank because of its location and the fact that it offered you a low minimum balance with no monthly fee.

But not all banks are alike. That's why it's important to shop around if you're looking for a place to deposit your hard-earned dollars. Knowing what to look for can help you save money on what it costs to bank.

Banking Choices

You shouldn't limit yourself to just banks if you're simply looking for a checking or savings account. There are other financial institutions that provide similar services.

1. Commercial Banks
2. Saving and Loans
3. Credit Unions
4. Private Banks
5. Mutual Funds
6. Brokerage Houses
7. Insurance Companies

What Do You Need?

Here are some things to consider:

- What kinds of transactions will you need to make?
- How much money will you keep in the account?
- How many checks do you plan to write a month?
- How much are you willing to deposit to have access to more services?
- What time of day do you normally go to the bank?
- Do you prefer the convenience of an ATM or would you rather bank in person?

Features to Compare

When shopping for a checking or savings account, it's important that you compare features. It pays to know the difference between accounts.

Look closely for some of the following features:

1) Services
- What selection of products does the bank offer?
- Holiday club savings.
- Senior accounts.
- Checking, savings, or money market accounts.
- Investment services.
- Insurance products.
- What is your perception of service at the bank?
- Do the bank's employees seem friendly and helpful?
- Do you feel comfortable in the bank's surroundings?

2) Convenience
- Does the bank offer online services?
- How many branches does the bank have?
- How close is their nearest branch to your home? Your work?
- How many ATMs are in the bank's network?
- Are the bank's hours flexible enough for your schedule?
- Are their branches co-located in convenient spots like supermarkets or malls?

3) Size
- Do you want a small neighborhood bank where you'll more likely to be remembered?
- Are you willing to forego lower costs for better service?
- In terms of service, how do large banks compare with smaller ones in your local area?

4) Fees
- Are there flat fees for checking?
- Is there a fee if your balance drops below a preset level?
- ATM usage fee? Fee for using ATMs outside of the banks network?
- Balance inquiries.
- Per-check fees.
- Overdraft protection.
- Bounced checks.
- Canceled checks.
- Closing your account.
- Are any fees waived if you set up a direct deposit with your paycheck?
- Are any fees reduced if you open up more than one account with the same bank?
- Loan application processing.
- Safe deposit box rental.
- Debit card use.
- Check fees.

5) Rates
- What is the current interest rate?
- Are rates based on the level of your account balance?
- How often is interest compounded?
- What balance is required to start earning interest?
- When does the interest begin accruing? When you deposit money or when the check clears?
- What is the Annual Percentage Yield (APY)?

6) FDIC
Look for banks that are insured by the Federal Deposit Insurance Corporation (FDIC). The government protects your money for up to $100,000 in losses. Ask the bank what FDIC rules will apply to your account. Look for a bank's Member FDIC sticker.

Opening an Account
What you need to open an account:

1) Valid Identification.
You should bring a valid I.D. such as a driver's license, social security card, birth certificate, or passport.

2) Social Security Number.
Banks require you to provide a social security number to open an account -- for tax purposes.

3) Signature Card.
You must sign a signature card so the bank can identify your handwriting on future checks, deposits, or other banking services.

4) Application.
The application tells the bank what type of account you're going to open (joint, individual, etc.). You'll have to list important information such as full name, address, phone number and other personal information.

5) Initial Deposit.
Most banks require you to make a minimum intial deposit to open a bank account.

6) Direct Deposit Form.
You have to fill this form out if you want your paycheck to be automatically deposited to your new account. You'll also have to notify the company for which you work -- so they can complete the appropriate paperwork.

Virtual Banking and the Bottom Line

Title: *Virtual Banking and the Bottom Line*
Author: *Elizabeth Sikorovsky and Liam Case, Tradespeak*
Abstract: *Virtual banks are forsaking a physical presence to challenge traditional retail banks. Will their strategy lead to long-term success?*

Virtual Banking and the Bottom Line

Issue

No Bricks, No Mortar, No Problem?
It would be a banking executive's dream to somehow magically eliminate the cost of maintaining branches. Real estate, maintenance, and personnel expenses add up quickly, and retail banks have been trying to mitigate branch network costs in recent years. Spurred by consolidation in the banking industry, many banks have closed branches to improve earnings.

Virtual banks are taking a more radical approach, aiming to operate with no branches at all. Operating branch-free lowers costs, which frees the bank to offer better rates and lower fees. The underlying premise is that the higher rates and lower fees offer a sustainable competitive advantage that will ultimately lead to profitability.

But for virtual banks, lower costs don't come cheap. There are trade-offs involved. Lacking the brick and mortar facilities of traditional banks, virtual banks can't establish profitable lending relationships with small businesses and retail customers. Because they can't establish these relationships, virtual banks are forced to invest their assets in lower-yielding capital market instruments. Virtual banks also forego the brand recognition and marketing benefits that a branch's physical presence provides.

Furthermore, virtual banks so far have shifted their costs rather than cut them. Although virtual banks' facilities and personnel costs are low, their marketing costs are sky-high relative to their offline counterparts. By forgoing a physical presence, virtual banks have knocked out an important means of differentiating themselves. Branches strengthen brand recognition, and offer opportunities for banks to differentiate themselves on the basis of service and convenience. Virtual banks attempt to make up for this lost value by spending more on advertising.

Some argue that growing the consumer base is the answer to virtual banks' profitability problems. It is not. Economies of scale will lower per-customer service costs, but the savings won't solve the bigger problem of narrower interest margins caused by high deposit rates and low investment returns. Others argue that virtual banks will solve their problems by originating more loans online and investing in relatively fewer secondary market instruments. But there's no guarantee that originating loans online will help margins either. As more loan originators become web-enabled, an undesirable rate game will follow as consumers use aggregator sites.

Differentiation through service and convenience, not just price, is the surest road to profitability. Such differentiation requires both an online and an offline presence, welcome news to traditional offline players. The trick for both virtual and offline players will be to offer the best mix of online and offline conveniences, plus competitive rates, and deliver it all at a price the bank can afford.

Analysis

Virtual Banks Compete on Price

The virtual banks in the sample – Net.B@nk, Telebank, Security First, and First Internet Bank of Indiana – offer 87 percent higher checking rates, 29 percent higher money market rates, and on average 27 percent higher CD rates than the banks in the offline sample. Because the three largest virtual banks operate under thrift charters, we compared them to five offline banks with thrift charters. For more details on the sample banks, please refer to the Approach at the end of the piece.

For virtual banks, higher rates aren't just an added benefit for customers; they are the fundamental differentiating factor used to attract deposits. Without the high rates, there would be no compelling reason for a customer to choose a virtual bank over a physical bank with online banking services.

By focusing on rates as their key advantage over the competition, virtual banks trap themselves into paying more for deposits. They must continue to pay high rates or they risk losing their clientele – clientele

Table 1: Deposit Rate Comparison, 2Q99

Product	Virtual Banks	Offline Banks	National Average of Deposit Institutions
Interest Checking	2.92%	1.56%	1.53%
Money Market	4.84%	3.76%	2.46%
3 Month CD	4.98%	3.86%	3.80%
6 Month CD	5.48%	4.28%	4.24%
1 Year CD	5.65%	4.49%	4.48%
3 Year CD	6.06%	4.89%	4.71%

Source: Thrift Financial Reports; Company Data; Bank Rate Monitor

who tend to be more price-sensitive, given that they were attracted by the value proposition in the first place.

Higher RoA for Traditional Players
A comparative analysis shows that traditional banks are making a return on assets much higher than virtual banks. Traditional players realize 1.07 percent annualized return on assets compared to virtual banks' -0.02 percent annualized return. It should be noted that our virtual bank sample has a wide range from Telebank's 0.44 percent ROA to Security First's -9.14 percent ROA. The national average for all deposit institutions is 1.23 percent, and 0.98 percent for all thrift institutions.

Arguably, a -0.02 percent return on assets is no cause for alarm for a group of pure Internet companies. For most pure plays today, operating in the red is the norm. But the business model, when examined closely, tells a different story. Virtual banks' strategy of offering high rates while maintaining no branches actually leads to lower profits compared to traditional players, in both the short and long run. Their lower profitability stems from three strategic vulnerabilities inherent in the virtual banking model:

- Because their fundamental value proposition is to offer customers higher rates for less service, virtual banks by definition must compete more aggressively on price.
- Most of the loan/mortgage origination process has not yet been successfully brought online. Thus, virtual players are limited in their ability to originate loans, an important source of profit. Even when virtual banks do begin to originate more loans online, their lack of physical presence will limit their ability to differentiate themselves, making them more vulnerable to price pressure than banks that originate loans offline.
- By foregoing bricks and mortar, virtual banks surrender an important means of differentiating themselves. They are forced to launch expensive advertising campaigns to gain brand recognition, nullifying the benefits of cutting facilities and personnel costs.

Growth alone will not solve these problems. Virtual banks cannot eliminate the vulnerabilities of their business model simply by achieving economies of scale, or creating online loan origination systems. Instead, virtual banks must create new ways of differentiating their products through service and convenience.

Creating compelling services and conveniences will require some offline presence, in the form of kiosk-like "micro-branches", or the right number of strategically located brick and mortar branches serving a larger number of customers per branch than traditional competitors.

Virtual Banks Struggle to Invest Profitably
Virtual banks pay relatively higher rates to attract customers; at the same time, their investments earn relatively less. Virtual players make an average of 6.47 percent annual percentage rate (APR) on every interest-earning asset dollar, compared to 7.14 percent APR made by traditional banks. In total, virtual players realize $0.19 less of net interest income per dollar of

interest revenue, compared to traditional banks.

This difference can be traced to virtual players' lack of branches. Without a physical infrastructure, these banks are severely constrained in their ability to originate loans, lowering their returns. Notably, virtual banks earn 71 basis points less than offline players in consumer mortgage loans, with offline banks earning 7.59 percent and virtual banks 6.88 percent. Virtual banks typically buy mortgage loans on the secondary market rather than originating them on their own. As a result, the virtual banks lose associated underwriting and servicing fees, and take on additional credit risk by allowing outside parties to evaluate the credit. While virtual banks do decrease credit risk through geographic diversification, it isn't enough to offset the loss.

Because they originate fewer loans themselves, virtual banks invest relatively more assets in mortgage pool securities than do their offline competitors, 35.4 percent compared to 7.6 percent. Mortgage pool securities are less profitable than mortgage loans. Virtual banks do earn relatively better returns on mortgage pool securities compared to offline players. However, the higher return suggests that virtual banks are also taking on more risk in the process.

Figure 1: Traditional Banks January - June 1999: Assets and Interest Income

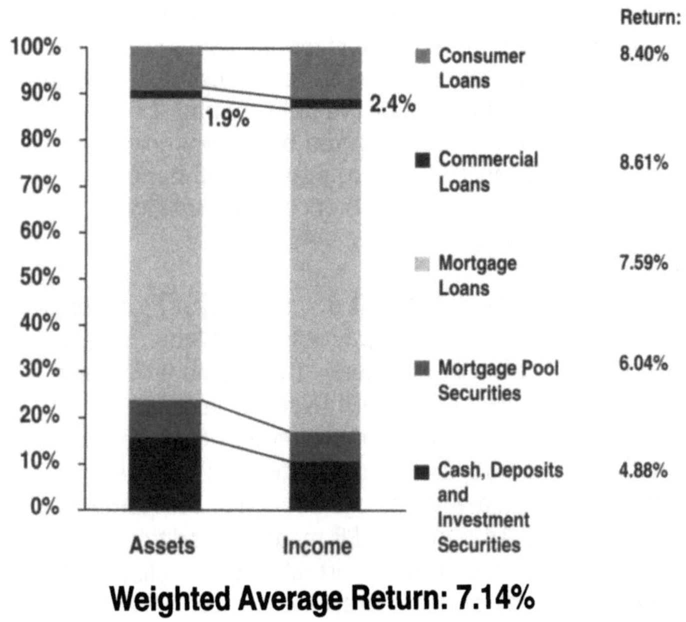

Source: Thrift Financial Reports; Mainspring

Virtual banks earn 58 basis points less on cash, deposits, and investment securities than do offline players. This inequality may even out in the long run as virtual banks and their investment decisions mature.

Predicted Savings Can't Erase Shortfalls Elsewhere
Virtual banks hope their slimmer interest income margins will be offset by lower non-interest expenses – specifically lower personnel and equipment costs. So far, the savings haven't been enough.

Figure 2: Online Banks January - June 1999: Assets and Interest Income

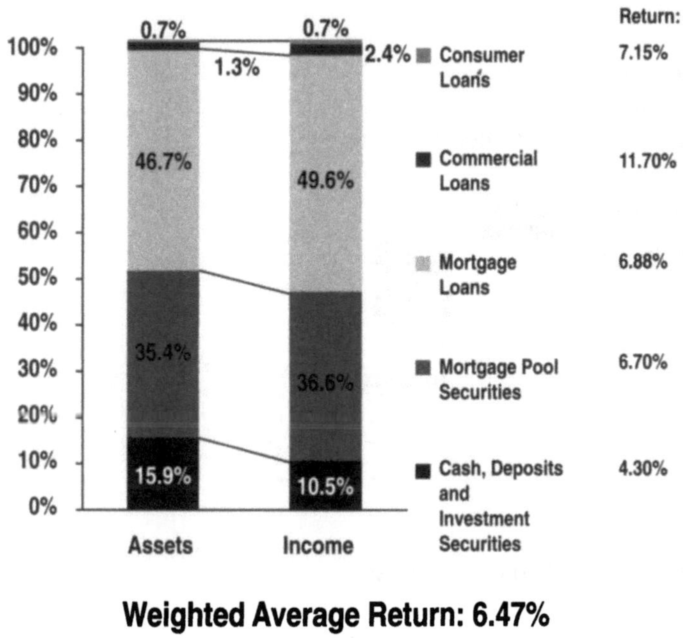

Source: Thrift Financial Reports; Mainspring

The figures 1 and 2 compare the income statements of traditional and virtual banks. Traditional banks have a wider interest margin than virtual banks, earning $0.44 in net interest income per interest revenue dollar as compared to $0.25 earned by virtual banks. Traditional banks also earn more in non-interest revenue, generating $0.095 for every interest revenue dollar, as compared to $0.085 generated by virtual banks.

Virtual banks do have lower personnel costs, spending about $0.065 per interest revenue dollar, compared to $0.165 for the offline banks. Virtual banks also saved two cents on office and equipment expense, with costs of $0.04 against $0.06 for offline banks. These cost savings should improve further through economies of scale as the virtual banks mature.

However, sky-high marketing costs more than nullify virtual banks' advantages in personnel, office, and equipment. Virtual banks' marketing costs are currently almost 10 times more than their offline counterparts, translating into a $0.125 cost difference for every dollar of interest revenue.

Virtual banks also exceed their traditional counterparts in other non-interest expenses, which includes everything from travel to data processing costs, with a cost of $0.085 per interest revenue dollar compared to $0.05 spent by traditional banks.

Taken together, virtual banks' non-interest expenses are 14 percent higher than those of their traditional counterparts: $0.33 per interest revenue dollar, compared to $0.29. Traditional banks realize a pre-tax profit of $0.245 on every dollar of interest revenue, compared to only $0.005 for virtual banks.

Figure 3: Income Statement, Traditional Banks, January - June 1999

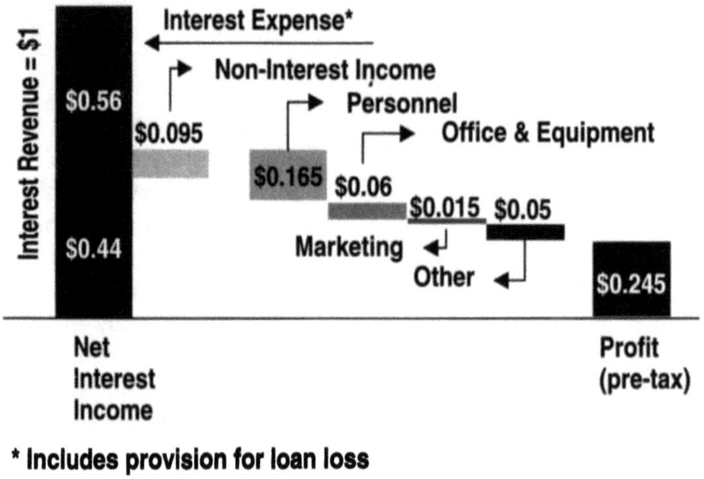

Source: Thrift Financial Reports; Mainspring

Figure 4: Income Statement, Virtual Banks, January - June 1999

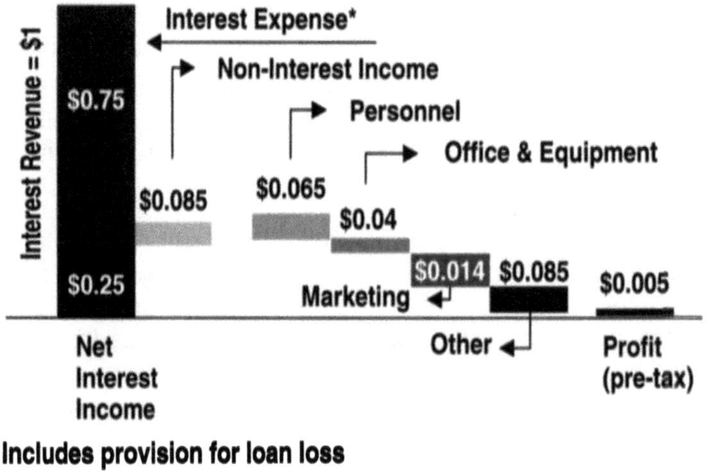

Source: Thrift Financial Reports; Mainspring

Virtual banks' high non-interest expenses can be attributed to their scale-tipping marketing costs. Virtual banks have comparatively higher advertising costs not only because they're new, but because they lack branches. Physical presence has marketing value, in that it helps establish a recognizable brand, and can promote an image of convenience and trust. Virtual banks have to spend more on advertising because they lack the marketing value that physical presence brings. Thus, eliminating a physical infrastructure may have the effect of shifting some costs (i.e. real estate to marketing), rather than cutting them. People don't walk by a Net.B@nk branch every day on their way to work.

Is Growth the Answer for Virtual Players?
Still, some argue that virtual banks' business model will only really take hold after virtual banks gather enough customers to gain economies of scale, and have established their brand enough to cut their marketing budgets. But even a significant drop in non-interest expenses, including marketing, would not be enough to pull virtual banks ahead of their offline competitors.

Consider this scenario: even if virtual banks eliminated their personnel and office and equipment expenses completely, and then squeezed their other non-interest expenses (excluding marketing) to the offline average of $0.05, they would still be less profitable than traditional banks by ten cents per dollar of interest revenue. Virtual banks would earn a profit of $0.145, compared to $0.245 for the offline banks. The $0.10 difference would remain because virtual banks pay so much for marketing. Virtual banks would have to cut marketing expenses by 70 percent, in order to realize the same net income as their offline competitors.

Telebank's experience bears this out. As the oldest and largest of the four sample virtual banks, Telebank has personnel and office & equipment expenses of $0.03 and $0.01 respectively, significantly lower than the virtual banking group as a whole. Telebank also enjoys lower non-interest expenses of $0.05, equal to the offline banks. But no end is in sight for Telebank's large marketing expenditures, taking $0.11 out of every interest revenue dollar. Telebank is still experiencing a return on equity far below that of its offline competitors: 4.02 percent vs. 11.64 percent. In order to achieve a return on equity of 11.64 percent, Telebank would need to have more than 375,000 accounts (over five times their second quarter total). This projection assumes continuing decreases in costs due to economies of scale, similar to what the bank has recently experienced – and the necessary increase in equity.

Overall, virtual banks' return on equity is even worse than Telebank's. Even when virtual banks' marketing expenses and equity/asset ratios are set identical to those of offline banks, virtual banks' return on equity still lags behind offline players by almost two percentage points. Virtual banks would realize an ROE of 9.97 percent, compared to offline banks current 11.64 percent, as shownin table 2.

Table 2: Financial Ratios for Offline and Online Banks, January - June 1999

	Offline Banks	Virtual Banks
Equity/Assets	9.2%	13.4%
RoE (annualized)	11.64%	-0.18%
RoE if Equity = 9.2% of Assets		-0.26%
RoE if Marketing = 1 percent of Interest Income		6.76%
RoE with both Equity and Marketing Normalized		9.97%

Source: Thrift Financial Reports; Company Data; Bank Rate Monitor

Advice

Virtual Banks Have Their Work Cut Out for Them

The virtual bank business model, interesting at first glance, looks significantly less attractive upon closer examination. Virtual banks cannot be as profitable as their offline brethren until they address certain fundamental issues. The first issue is to widen the slim interest margin. Originating more loans online is not the answer. True, virtual banks will originate more loans themselves. But pricing competition will result in the commoditization of online loan origination, eliminating the possibility for fatter margins.

Virtual banks' ability to up their return on equity does not depend on how low they can push their costs. Instead, it depends on their ability to differentiate themselves in a market characterized by strong price competition. Thus, differentiation on something other than price will be a key determinant in whether or not virtual banks can develop a sustainable advantage.

Offering a wider range of financial services and products, building physical branches or kiosks – as Salem Five has planned to do to support its online banking efforts – or even providing non-FSI products through partnerships are ways to differentiate and improve the value proposition for customers who might otherwise demand better rates. Brick and mortar facilities also provide shelter against loan rate competition, by creating an opportunity for loan officers to build relationships built on service, not just rates. A physical presence also boosts marketing efforts, an area where virtual banks are already spending a lot of money.

Who Has the Advantage?

If banks must battle for customers in both environments, who has the advantage: traditional offline or virtual banks? Virtual players have the advantage in infrastructure costs and agility; traditional players have the advantage in brand recognition and customer relationships. Virtual banks have an infrastructure cost advantage over traditional players, even in the offline world. Although a new branch infrastructure built from the ground up is expensive, scaling back an existing infrastructure (as traditional offline players must do) can cost even more. Virtual banks also have the advantage of being able to build kiosks, or micro-branches, solely devoted to loan originations.

Virtual banks can leapfrog existing banks by choosing the optimal branch to customer ratio, and then building or buying the infrastructure accordingly. Virtual banks can choose the size and location of their branches to maximize visibility and convenience, while still leaving much of the service offered to the lower cost web channel. They can also buy branches as traditional banks divest them. Traditional offline banks start with an existing infrastructure, built for customers who do not use the Internet.

In addition to lower costs, virtual banks have greater agility. They do not have the existing customer base that their offline competitors have. Nor must they protect the integrity of a brand that has taken decades to build. These advantages leave virtual banks more free to innovate. Virtual banks are also unencumbered by legacy systems and legacy modes of thinking.

Traditional players have an advantage in that they are more familiar to customers today. The large traditional banks have something that the virtual banks are spending millions of dollars to get right now: brand recognition. A virtual bank must become a known quantity before it can expect to lure a significant number of customers away from traditional banks. Virtual banks will have to continue aggressively advertising to boost recognition and loyalty among customers.

Traditional players with existing physical infrastructures have an opportunity to excel. Their biggest challenge will be in choosing which of their offline assets they will use to support their online banking, and which they'll keep separate or shut down.

Approach

Virtual Banks vs. Traditional Banks
To analyze the economics of different banks, we compared the financial statements of four virtual banks with those of five comparably sized brick and mortar banks.

The traditional banks – Brooklyn Federal Savings Bank, First Federal Lincoln Bank, Harbor Federal Savings Bank, Home Savings Bank, and York Federal Savings and Loan Association – range in size from approximately $200 million in assets to $1.4 billion. They all operate under thrift charters.

The virtual banks that form the basis of this analysis – Net.B@nk, Telebank, Security First Network Bank, and First Internet Bank of Indiana – vary in size from under $50 million in assets to over $3 billion, as of the end of the second quarter 1999. Of all the institutions in the analysis, only First Internet Bank of Indiana operates with a commercial bank charter; the rest are thrifts.

Table 3: Offline Banks: Assets, Equity, Accounts, and Branches, 2Q99

Offline Banks	**Assets ($K)**	**Equity ($K)**	**Accounts** ***Branch***	**Accounts/ Branch**
Brooklyn Federal Savings Bank	$209,407	$20,725	21,753 ***3***	7,251
First Federal Lincoln Bank	$1,216,005	$95,010	92,736 ***56***	1,656
Harbor Federal Savings Bank	$1,432,963	$140,929	111,600 ***31***	3,600
Home Savings Bank	$744,860	$70,305	86,192 ***16***	5,387
York Federal Savings and Loan Association	$1,356,502	$93,649	133,100 ***22***	6,050
Average	$991,121	$84,124	122,598 ***25.6***	4,789

Source: Thrift Financial Reports; Company Data; Bank Rate Monitor

Table 4: Virtual Banks: Assets, Equity, and Accounts, 2Q99

Virtual Banks	**Assets** ($K)	**Equity** ($K)	**Accounts**
First Internet Bank of Indiana	$47,298	$12,802	1,387
Net.B@nk	$801,034	$128,077	33,653
Security First	$168,937	$27,242	21,526
Telebank	$3,191,712	$298,765	69,308
Average	$1,052,245	$116,729	31,469

Source: Thrift Financial Reports; Company Data; Bank Rate Monitor

Virtual banks are more likely to be thrifts, because thrift charters can be acquired more quickly. However, some virtual players are registered as commercial banks, such as USABancshares.com, USAccess, and WingspanBank, all of whom operate under a previously existing charter.

A thrift charter has two important effects on the way the bank invests its money. First, federal regulations restrict the amount of commercial loans those banks can keep on their balance sheet, limiting a potentially lucrative area of interest income. Second, the banks enjoy certain tax advantages if over 60 percent of assets are invested in mortgage loans or mortgage derivatives. Thrifts therefore tend to have mainly retail customers, and only limited small business activity.

Banking in the Future

Title: *Banking in the Future*
Author: *Bank Technologies Group*
Abstract: *A paper from bankers for bankers about using the Internet as a new channel.*

Banking in the Future

Electronic Banking and the Internet

The development of electronic financial markets is moving at the speed of light, both figuratively and literally. One year ago there were fewer than one hundred financial firms, including banks, brokerage houses, savings and loans, and credit unions with representation on the Internet. Today one can find five hundred banks with home pages. Furthermore, a year ago these home pages were really nothing more than advertising (i.e., electronic billboards).

A year later, these internet banking sites process customer service inquiries, allow transactions from one account to another, take loan applications, open new accounts and enable customers to trade e-mail messages with the President. At least one, First Union, has become the first bank in the world to offer international trade finance to its customers through the Internet. Some provide commercial services and a few are basically full service banks or rushing to get there. In addition to the rapid deployment of electronic bank "branches," an entire financial community has suddenly appeared, offering most any financial service a consumer could want. New forms of money are being coined to pay for transactions, and untraceable payment protocols are being tested and marketed. The number of seminars and trade association meetings being held, as well as their attendance, is further indication that the march to the Internet is really a "charge."

As fewer and fewer bank customers visit bank lobbies and drive-in lanes, more and more are using ATM's, automated call response, home banking terminals, and/or the Internet to do their financial business. This "charge" to the Internet is not a passing fad to be ignored. It is an aggressive, offensive strategy to capture a large share of the financial services market. Supported by speed of light telecommunications, powerful, inexpensive computers and secure encryption technologies, our competitors are racing into our best customers home via the Information Super Highway (that is, the Internet).

Furthermore, the charge to the Internet is bringing new competitors into our markets. The Internet consumer has choices of 556 banks from 35 countries with whom to do business. The Internet charge is driven by sound economic and consumer service issues as well (mainly delivery cost and customer convenience). We, as bankers, face a choice: we can spend $750,000 and a year to build a branch that might reach 100,000 potential customers, or we can reach out and touch 10,000,000 potential customers for less than $350,000, and do it in half the time. Not only that, but a traditional branch of brick and mortar would do nothing for our existing customers; whereas, the convenience for all our customers would be greatly enhanced by an electronic, 24 hour branch. The cost of staffing, training and operating an electronic branch is much less as well. Perhaps the most enticing temptation of all, is the profile of Internet customers: educated, professional, affluent.

An additional factor that has engendered the rush to Internet banking is the continuing growth of near banks and non-banks, both foreign and domestic. The majority of bankers today will tell you that their greatest competition is not from other banks, but from credit unions, brokerage houses, (the Merrill Lynches of the world,) with the majority of the "lost" deposit dollars going to the brokerage houses. "Banks today hold only 20% of household financial assets, versus 34% twenty years ago; they have 30% of business deposits versus 42% only seven years ago. Non-bank credit card providers have gained inroads against banks, holding a 25% market share versus 5% in 1986." What will happen when these near banks and non-banks are more convenient than we are? It is basically convenience (i.e., location) that has allowed banks to survive this long . There is little doubt, however, that this principal advantage is disappearing as the use of the Internet for financial products continues its rapid development.

The last factor driving banks to the Internet is fear: Fear that some major non-banking entity, (such as Microsoft) will form alliances with a brokerage house or credit union to offer financial services world-wide. When that, or something similar to it happens, we had better be at least one step ahead.

Regulatory Risk

In Banking 101, we were all taught that the Federal Reserve and the Federal Deposit Insurance Corporation were the "...money stewards of the society." However, there is a large body of research that shows the continuing transfer of deposits to entities that are all but outside regulatory oversight. This trend may accelerate as Cybercash, CheckFree, Ecash, NetCash, and other near moneys and payment methods proliferate on the Internet. What percentage of the money supply will the regulators be able to control in the future? Will money, in the electronic age, run to the country with the most advantageous banking system? Will "real" money be driven out by virtual money? Are we facing another era like existed in the United States from 1836 to 1863 when there was no central banking authority controlling the money supply and private companies (i.e., state banks) issued their own bank notes? Will banks be reduced to repositories of Federal Reserve Notes backing privately issued cyber-cash?

Ignoring these monetary issues for the moment, how are banks to respond to the immediate challenges of banking on the Information Super Highway? A review of the Comptroller of the Currency Handbook reveals no policy or examination procedures as it pertains to home banking, not to mention the Internet. Is Internet access to banks transactions simply another form of communicating with one's bank?

Additional questions pertaining to the regulators' role in the "electronic branch" centers around current regulations. How does one apply CRA to a virtual bank with a nation-wide customer base? How do state regulations apply to virtual banks? Which state regulations apply, the state in which the bank is domiciled or where the customer resides? What if the customers are world-wide? What if one wanted to charter a national bank to serve only customers residing outside the United States? In effect, there are numerous regulatory issues evolving from the concept of electronic banking that are not addressed by current law, regulations or examination practices.

Perhaps the biggest issue, at least to the FDIC, is security. What standards will or should be required. Are current security and encryption methods adequate? (After all, two University of Purdue students broke the code used by the FRB's wire encryption system in eighteen minutes in an assigned exercise.) If the government continues to bar the export of encryption technology, will American banks be unable to compete on the World Wide Web, the fastest growing segment of the Internet?

Institution Risk

The ease with which banks and other financial entities may begin offering financial products on the Internet is another cause for concern. Will the rush to compete be a rush to disaster? Which are more dangerous, large banks with sophisticated systems putting huge sums at risk on the Internet, or small, unsophisticated banks risking small sums? Do imitators of Citibank or Security First Network Bank pose the larger problem?

What impact will the first major Internet bank failures have? Will the regulatory bodies become so restrictive as it pertains to the requirements of entering into electronic banking that they may slow the process, giving the advantage to non-banks and foreign banks?

Conclusion: The Future is Now

Electronic banking is a world wide phenomenon engendered by the revolution in computers and telecommunications. Just as most of us thought the Industrial Revolution was over, we find it has just begun. To survive against non-banks and foreign banks, we in the banking industry must adapt to the electronics age, and do it now. Consumers demand it. Economics drives it. Banks must exploit it.

Are Banks making the Most of it?

Title: *Are Banks making the Most of it?*
Author: *Credo (www.credogrp.com)*
Abstract: *A paper by the Credo group electronic banking and its development.*

Biography: Credo Group is a leading provider of e-banking solutions worldwide. Its FONTIS Internet and PC banking system has been implemented in over fourty countries. The company is part of Midas-Kapiti International (MKI).

E-BANKING. ARE BANKS MAKING THE MOST OF IT?

Banks have to grapple with how to gain a prominent place in the world of electronic commerce. At the outset, many saw the Internet as merely another delivery channel, bringing greater flexibility, reduced costs (no need to install software at all of those end-users) and security risks. However, in some organisations at least, the thinking has moved on. The real question is how to make the most innovative use of the technology and how to ensure that others do not beat you to it.

After the initial setting up of a web site for marketing purposes, banks started to replicate their existing electronic banking services using browser technology, with either HTML or Java. A couple of market surveys suggested that this form of banking would not only take off in the retail banking sector but would also come to dominate for corporate banking. Figures were bandied around, which showed a huge market value by 2000.

In reality, after initial sorties, banks started to take a more cautious approach. Indeed, a few came close to launching Internet-based banking services for their corporates and then went no further. Some of the strategies looked as though they had been technology led rather than business led. Lessons learned in other areas of banking had been forgotten in the rush to be state-of-the-art.

'It is not acceptable just to port existing functionality and products onto the Internet', says Tony Nicholls, corporate banking electronic business services manager at Barclays. 'Corporates do not exist to be banking customers. They are focused on buying and selling, with banks increasingly becoming just another service provider. The Internet provides a real opportunity to surprise our customers. Many do not know what they want yet, so think laterally', he advises. 'One major push will certainly be for more of everything - more information, faster communications, instant query resolution, 24x7 for all relevant services.'

'Frustratingly, the likes of Barclays are still in the minority' says FONTIS Business Development Manager, Leslie Duckett. 'Many are still stuck in the traditional mind-set. They think a straight move from electronic banking to Internet banking will solve all their problems. It is just not true.' To aid the move to a more sophisticated approach, Credo has developed a Secure Instruction Designer component for the web-enabled version of FONTIS. This is Java-based and allows banks - or Credo on their behalf - to develop any type of instruction with full validation capabilities. It can be used for traditional banking services as well as allowing banks to encapsulate broader e-commerce services.

At the simplest level, one bank which Credo is working with sponsors cricket and wants to be able to sell tickets via its Internet-based retail banking service. At a more complex level, some large financial services conglomerates are trying to use the Internet to offer a seamless set of services from the different entities within the group. At the same time, Internet technology should allow highly tailored marketing, with services and messages adapted according to the profile and habits of the customer. Barclays Nicholls says the Internet should allow banks to customise for a market of one.

In the retail arena, another area that is certain to be a hot topic over the next few years is electronic bill presentment and payment. The ability to view and settle bills on-line is already taking off in the US, where portal providers, telecoms companies and banks are battling to gain a slice of the market. Consumers can link directly to a biller but it is much more convenient to go to a single location and be presented with bills from various sources. In the US, the billers are being wooed from all sides. A commercial bank might support the biller, providing an electronic platform for delivery of bills to end-users. But that provider might equally be Yahoo! or a PTT.

It is likely that the electronic business-to-business arena will be another battlefield. Many companies have put in large ERP systems and would like to finally move to an electronic environment for purchasing and supply management - the final realisation of EDI. Corporates want straight-though processing for ordering, receiving product statements and other information, invoicing, and payment. Small to medium-sized organisations may well want browser-based interfaces into the electronic infrastructure, others will want integration with their ERP systems. Banks can seek to provide this sort of infrastructure, clearly working with technology partners as well as with components that are increasingly being offered more or less off-the-shelf. Alternatively, they can sit back and watch while corporates put in place these sorts of infrastructures themselves or use non-banking third parties. Banks could be merely left to handle whatever netted payments are required at the end of the process.

Speaking at a recent conference, Chris Robinson, Global Solutions Manager in Citibank's Cash Management Group, concluded: 'The market is evolving rapidly and opportunities are plentiful so be prepared to place some bets and be bold. Those opportunities might be plentiful, but they are also only temporary. In their favour, banks have the trust and confidence of their customers, as well as strong existing relationships. However, banks need to rapidly gain new skills as well: speed of action, creative thinking and the ability to harness and make full use of new technology as an enabler for well thought-out business strategies.'

ELECTRONIC DELIVERY CHANNELS IN BANKING

Competition and changes in technology and lifestyles have changed the face of banking. With the traditional branch based business on the decline, banks are now seeking alternative ways to provide and differentiate their services. Customers, both corporate and retail, are no longer willing to queue in banks, or wait on phones, for the most basic of services. They demand and expect to be able to transact their financial dealings where and when they wish.

With the number of business and home computers increasing every year, the electronic delivery of banking services is becoming the ideal way for banks to meet their clients' expectations. Initially provided by only the larger banks to their corporate clients, every bank realizes that the ability to provide electronic banking services is essential to its survival. Retail clients also becoming more demanding and it is estimated that over 40 million retail customers will be availing of financial services through their PCs by the year 2000.

One company that is doing a lot of work in the area of electronic banking is Midas-Kapiti International.

According to Richard Williams, Branch Manager at MKI Madrid, "We have been providing a quality service to our clients in Spain for a number of years. It is our job to make sure that we can provide them with the products and services they require. We identified Electronic Banking as a major growth area two years ago and introduced FONTIS, one of the most popular remote banking solutions to our range of products. FONTIS has been installed in banks throughout the world and is helping banks to provide their clients with a complete range of banking services".

At its most basic, electronic banking for retail customers means providing up to date account information. A funds transfer facility should also be offered to clients. Some banks restrict this service to moving money from one account to another e.g. current account to mortgage account for retail customers. This can however be expanded to give clients the ability to pay all household bills e.g. electricity, telephone, gas.

While this level of functionality may be sufficient for banks to provide to their corporate clients as an initial service, it should not be a long-term strategy. Business users have more sophisticated needs so the bank must ensure that they provide not only the functionality, but also the choice of technology to meets these needs.

Balance and Transaction reporting is the basic requirement of all business users, but they must be able to receive a complete financial complete financial status report. As clients often have relationships with other financial institutions many banks are offering a multi-bank reporting service. This feature should give corporate clients the ability to assess their overall financial position by grouping accounts in different scenarios, comparing their values with respect to different currencies. From January 1999 banks will have to offer a further service to their clients by reporting balances in both the currency of the account and in the new euro currency.

The area of Payment preparation for corporate clients is also more intricate. While the business users have the same basic needs to transfer money from one account to another, there can be currency implications. Banks must be able to offer cross currency payment facilities in conjunction with an up-to-date treasury rate service. International payments and formats must also be supported. Payment import and payroll facilities are also being introduced to reduce errors and time for both bank and client users. Many banks are finding the ability to transfer files is an essential service for corporate users. It enables them to quickly and securely transfer files to the bank rather than relying on outdated methods such as files, tapes and disks.

Trade Finance is also an area that is traditionally fraught with delays, errors and amendments. For this reason banks are now offering their international clients the ability to prepare electronic Letter of Credit (LC) Instructions. Systems that can also keep the client up-to-date on the status of LC instructions are becoming increasingly popular.

The electronic delivery of payments, instructions and other files has advantages for the bank also. Firstly it ensures that only complete, secure and authorised payments and LCs are sent to the bank. Also by receiving instructions electronically, the bank can adopt "straight-through-processing" making the processing of payments and other instructions even more error-free and efficient.

As well as deciding on the functionality they wish to provide, the Bank must then choose the type of customer they wish to target and select the technology best suited to them. Williams claims, "Many banks are now moving towards a global approach for electronic banking with each of their international branches providing a localised version of the same product. FONTIS was the only solution which could match this requirement and also give the bank the flexibility of either a Windows or Internet environment".

The general acceptance of the Internet makes it the ideal way for banks to reach a wider retail market without the expense of opening and resourcing additional branches. Internet products developed using Java technology do not have to be installed on the client PC but are temporarily downloaded by the client for the duration of their on-line session. This new technology makes it easier for the bank to support higher numbers of users as software upgrades and changes can be made centrally at the bank.

Many banks are also taking advantage of this technology for their corporate users, reaching a wider international market with their products. However due to the perceived problems with the openness of the Internet, many of the larger corporate clients prefer the traditional dial-up solutions, connecting to the bank only to send or receive information.

"The key to providing a successful electronic banking system is in listening to your target market, identifying what they want from a system and then selecting the system which provides this functionality through the technology they require" says Williams.
He continues, "I believe the Y2000 issue will see a lot of banks making quick decisions about electronic banking, either replacing old systems or introducing new ones. Such banks must take care that they don't buy without considering all aspects of the electronic banking project. While euro and Y2K compliance are essential the banks must also ensure that they are providing the right product for the right market".

BANKING THROUGH ELECTRONIC CHANNELS - SHOULD BANKS BUY OR DEVELOP ?

The electronic delivery of banking services is growing in popularity each year. Initially provided by only the larger banks, every bank now realizes that it must be able to provide even the most basic of services to their clients PCs to survive. It is estimated that over 40 million retail customers will be availing of financial services through their PCs by the year 2000.

Many of the larger banks with sufficient IT resources have opted to develop their own system. While this gives banks the freedom to develop a system that meets 100% of their needs, the time to develop such systems has often meant that such banks are late to market and loose out to those providing off-the-shelf solutions.

Bespoke developments by either a bank or software provider require at least a year and a half to reach completion. Whereas package based solutions often require a mere three to six months to implement. Once you consider these time differences it is not hard to see that sometimes it pays to compromise on some functional requirements in order to attain competitive advantage.

Gerry English, Sales & Marketing Director for Credo Group, one of the leading providers of remote banking solutions claims, "Many banks start in-house developments without first considering if they have the knowledge and skills necessary for such a niche development. This has meant both delays in developing the product and has means that a bank may end up with a package which functions correctly for the bank, but does not meet their clients needs".

English continues, "We have spent over a decade in this business learning how to anticipate and meet changes in technology and functional requirements. Our FONTIS product was the first Windows based system and also the first to offer both Internet and Windows based client services from a single system. How can anything but the largest of banks devote the required time, money and resources to match such a product".

Time to market is only one of the issues that should be considered when deciding between a vendor package and an in-house development. Functionality and user friendliness are extremely important for the success of any remote banking product.

When selecting a vendor system or defining a bespoke development, care must be taken to ensure that the system is as user friendly as possible for both bank and client staff. A system that is difficult to use and navigate will not be used to its full capabilities. The screens must be uncluttered and also make the best use of icons, buttons and drop down menus. Users must also be capable of exporting information, as well as viewing it on screen.

In the area of payments, banks must be able to define the information and format required for each payment type. Mandatory fields are essential to ensure that only complete payment instructions are sent by clients. Care must also be taken to ensure that any hardware or software security, especially in the area of payments, is effective but not intrusive.

Software by its nature is a dynamic medium so banks must ensure that the system they purchase or develop is able to adapt to incorporate new technologies and also new services. Without the flexibility to add new functionality when it is required, banks will find the life span of their investment severely curtailed.

According to Gerry English, "Some banks who opt for package solutions believe that they have to sacrifice flexibility and functionality to save both time and money. That should not be the case. We developed FONTIS with a series of Windows based tools to enable the bank to easily and quickly to develop their own screens and customize the standard ones".

One final factor, which must be taken in to consideration when selecting a suitable provider, especially in the area of Internet development, is financial stability. "Banks need to protect their investments and can not afford to purchase products from companies that may not be around to support or enhance their products", claims English. "We have identified a number of small providers who claim to have Internet banking systems. Unfortunately, as there are more vendors than customers in the marketplace, only the financially stable will survive".

Net Banking Benefits! Sheer Acceleration

Title: *Net Banking Benefits! Sheer Acceleration*
Author: *BankAway!*
Abstract: *Banks have realized that customer centricity, reduced costs and a sophisticated remote delivery infrastructure is the key to success in the self-service business. The Internet promises to emerge as the predominant delivery channel with unmatched benefits.*

Introduction

Given the exponential growth of the Internet in the past two years, it is increasingly apparent that banks and financial services firms will have to adjust to a rapidly changing competitive situation in which the Internet, Extranet and the Intranet infrastructure will be a key differentiator.

According to a Booz Allen study :

- Major banks plan to quadruple their investment in the Internet.
- 42% plan to have advanced sites offering complete interactive banking.
- The US, Canada and Australia account for 1005 of the present advanced sites.
- North America and Europe account for 79% of total current sites

Internet banking is still in the early stages, but the broad end user segments from the banker's point of view are quite clear.

- The individual consumer market
- The corporate customer market

The individual consumers
It has been estimated that by the year 2000 there will be 16 Million1 US households who will use Internet banking. More importantly these customers will be the most profit yielding customers representing 30% of retail banking profits. This high profitability will be driven by the attractive demographics and higher than average financial needs of Internet banking customers. The figure of 16 Million US households having Internet banking has been arrived at by taking into consideration the factors affecting consumer demand for Internet banking, computer ownership, Internet usage and consumer acceptance of electronic banking

The corporate customer market
Corporate Consumers execute and manage thousands of financial transactions daily; need multiple updates throughout the day; and have individuals in different locations who must access their account information--all of which can be accomplished quickly and cost-effectively through either the Internet or Extranets

and therefore corporate customers would opt for banking over the net. Following in the footsteps of retail banking, corporate banks are building a global presence on the Internet. This will also facilitate banks' global expansion and dramatically increase competition for the limited number of corporate customers. The US, Canada, and Australia, followed by Western Europe, are the most advanced in terms of on-line corporate banking capability, with Asia and Latin America in the earliest stages of development.

The Stakeholders

The stakeholders in the Internet Banking scenario analyzed from a very broad point of view are:

- Banks & Financial Institutions
- The retail customer
- The corporate bank customer

Banks

According to a study by Booz Allen, there is a huge perception gap between North American/ European banks and Japanese banks regarding the future of Internet banking. While North American and European banks foresee the Internet becoming an important banking services delivery channel within ten years, the Japanese banks expect traditional branches to remain the most important delivery channel.

Grant-Thornton Consultants conducted a survey on Internet banking of US banks in 1996. According to this survey, around 39% established the site to increase their client-base and to remain competitive with other banks.

Grant-Thornton Consultants also found that

- Around 59% of the banks had brought in new customers via their Web-site.
- Around 34% said that they had sold products and services to existing customers via their Web-site.
- Another 46% said that when people request information via our Web-site, the banks can track whether these requests result in sales.

All this clearly indicates that banks are bound to adopt the Internet as a delivery channel in the near future.

The major benefits to the banks offering Internet banking services are:

- Better branding
  The banks offering services on the net currently, are a very few so, those who would offer such services would be perceived as leaders in technology implementation. Therefore, they would enjoy a better brand image
- Reduced Transaction Costs
  This is in terms of the cost of delivering various banking products and services.
- Better responsiveness to the market
  Banks which enter early will be able to get a better handle on the services and products which the consumers need. Also, these banks will require less time to develop and deliver products to their customers
- Greater revenues
  The banks can use their web sites to sell their own as well as other third party products to their customers. The diverse range of products and services which can be sold through their web sites will lead to increased revenues for the banks.

The above benefits would draw in a greater number of customers as well as revenues for the bank.

However, with many of their profitable customers using Internet banking, banks could face some strategic risks. Chief among these risks will be the ability of a bank's customers to easily switch banks. These low switching costs could make it much harder for banks to hold on to some of their existing customers. In addition, two entirely new classes of competitors, Agents and Integrators will compete with banks for the control of Internet banking customers. Together, all these factors will make Internet banking a much more competitive environment.

Retail Customers
When an individual customer looks at banking via the Web, there are chiefly, five questions that come to his mind:

1. Can the bank offer a wide range of products and services over the Web?
2. Can I effect funds transfer between accounts?
3. Can I download my account history on a PFM software and handle my scheduling and planning of payments better?
4. Is it secure to transact over the Web?
5. Can I get access to other websites such as shopping malls and news-houses, etc., through the bank?

Most of the banks today are addressing these issues while launching banking on the Internet. The major benefits that the retail customers can expect is:

- Reduced Costs
  This is in terms of the cost of availing and using the various banking products and services.
- Convenience
  All the banking transactions can be performed from the comfort of the home or office or from the place a customer wants to.
- Speed
  The response of the medium is very fast, therefore customers can actually wait till the last minute before effecting a fund transfer.
- Funds Management
  Customers can download their history of different accounts and do a 'What-if' analysis on their own PC before affecting any transaction on the web. This will lead to better funds management.

Given the above benefits, who would not like to shop from his/her home?

Corporate customers
When any corporate customer looks at Internet banking, he/she has just one question in mind, "How can Internet banking benefit my business ?"
The benefits are quite a few, namely:

- Reduced Costs
  This is in terms of the cost of accessing and using the various banking products and services, applying for loans, L/Cs etc.
- Access to information
  Corporations will have easier access to information as, they can check on multiple accounts at the click of a button.
- Better cash management
  Internet banking lets the corporate

customers move cash between accounts to pay bills while they still earn as much interest as possible on their cash reserves. With PC-to-PC banking, cash-management decisions are transmitted from the business to the bank by computers. With up-to-the-minute information at their fingertips, business owners can make more timely decisions about how to manage their funds.

But despite these benefits, corporate bankers are slow in shifting to the Internet. This is because, the corporate banking market is quite different from the retail banking market.

- Corporations require multi-user environments often with different levels of authorisation w.r.t., transactions. Situations can arise where only the head office can fully authorise certain transactions centrally or where transactions require more than one authorisations. This set-up requires that data for a particular client be held centrally and accessed by their different offices.
- Corporate banking volumes are of a different scale when compared with retail banking. Corporate banking involves multiple account transfers and repetitive payments. It is therefore a distinct advantage to be able to prepare large volume transactions off-line before connecting to the bank.
- The lack of wholesale banks' aggressiveness compared with their retail counterparts can be attributed to the perception that Internet's growth has been more of a consumer phenomenon.
- In many cases either lack of understanding about security on the Internet, or lack of confidence in their understanding of the security issues is holding back many corporate banking units.
- No published standards are available which can be deployed for implementing corporate banking
- The existing corporates have diverse resource planning or automation systems within the organisation.

However, despite the slow start, banks have started launching interactive web sites for their corporate clients and it is expected to catch on and successfully evolve over the next couple of years.

The Total Pie

The benefits of Internet banking are complemented by the growing market size. The market size can be looked at from two aspects,

- By number of households
- By number of transactions

Since, most data available at present is of US retail consumers, the market size analysis here is only keeping in view US retail customers.

By number of households

Internet banking is a part of a larger market i.e., the on-line banking market. In March'97, the total number of on-line banking households in the US were 2.1 million. This includes those households who access to their bank accounts through the Internet as well as through a home based PC. This is only 6 % of households with PC banking facilities (34.8 million) in the US.

However, there are some interesting statistics from 1996 American Banker / Gallup Consumer Survey. A few details regarding this are displayed in the table below.

From the table below, it can be seen that around 18 million households is the potential market for on-line banking in the immediate future. Most of these households would be common to all the three on-line banking channels, PC homebanking, telephone and the Internet.

By number of transactions

The number of homebanking transactions processed in 1995 has been estimated to be 200 million. But, this is expected to grow to 1,400,000 million4 transactions by the year 2000. This translates into approximately 600% growth rate annually. Most of these transactions are expected to be bill payments and taking a charge of $4.0 (currently charged by Checkfree Corpn. per customer per month for 20 bills), the size of the transactions market in the year 2000 will be $280 billion. The size of the transactions market is an important measure from the banks' point of view as, their revenues and profits will be linked to the number of transactions processed. Banks are slowly realizing the immense benefits of Internet banking. At present, there are close to 1200 banks having websites. Most of these are in the US. In Europe, UK leads the pack with 35 banks with Internet presence, ahead of countries such as Spain, Italy, and France. However, only 35 banks in the US and 21 banks outside the US, provide Internet banking facilities (NetBanker.com/True Internet banks). A 17-country study of banking and financial institutions released by Ernst & Young shows that Canada and Australia lead the world in their commitment to

Table 1: PC owners with interest in on-line banking

S.no	**PC Owner Characteristics**	**% of households**	**Number of households** (in millions)
1	Have used on-line banking	6	2.1
2	Are very interested in on-line banking	23	8.0
3	Are somewhat interested in on-line banking	29	10.0
4	Total very or somewhat interested	52	18.0

Internet transactions. This study also points out that Internet and on-line banking facilities will increase exponentially in the coming years. In Canada for example, spending by banks for the development and maintenance of Internet sites will grow from $8.5 million in 1996 to $54.5 million in 1999. In contrast, bank branches will account for only 31 per cent of transactions in 1999, down from 45 per cent in 1996.

Internet Banking in Europe

Title: *Internet Banking in Europe*
Author: *ISPO – European Commission*
Abstract: *In the Internet banking area, Europe is often considered as lagging behind the US but this conception is untrue. Two major reasons can explain that phenomenon. The first one is that American banks put principally the stress on retail electronic banking while European banks have these last years mainly been busy with corporate electronic banking. The second one is that an important effort of standardisation has been realised in the USA regarding Internet banking interfaces while the technical solutions in Europe are almost as many as the number of banks offering Internet banking.*

Internet Banking in Europe: Key Issues

Introduction

Internet has become a new major communication medium and banking sector is perfectly aware of this phenomenon. An increasing number of banks now aims to develop Internet solutions for bank transactions. According to IBM Consulting Group, within 10 years, Internet will become the dominant delivery channel for financial services, even if currently Internet banking operations represent approximately 5% of transactions total volume both in the USA and in Europe.

The initial impetus for Internet banking has been the low transaction cost, compared to other channels. But cost advantages are not sufficient anymore to explain the current development of Internet banking. In the medium-term, Internet banking offers considerable scope for new services. The challenge for financial institutions is to find satisfactory answers to business issues of culture and processes and to create a win-win game which both satisfies banks and customers.

Developing Internet banking activities provide banks with numerous advantages. Firstly, Internet banking allows them to maintain the customer relationship or to make it evolving. The customer can actually be joined wherever an Internet access is possible (home, university, office, ...)and targeted customers are the wealthy, young, educated and profitable customers. Another advantage of Internet banking is that it increases revenues (product catalogues, proactive cross-sell). It also represents a competitive advantages through differentiation of banking services and image improvement. Finally, Internet banking allows a modular approach for cost effective development and is of course a lower entry cost gate than branch.

But banking activities migration towards internet can cause some concern too. The first one is the security. As we will seen it in a further section, security is a major aspect of Internet banking not only because of transactions interception and falsification but also because of the potential bank's private systems exposure. The second cause of concern is the rapid evolution of Internet and the fast obsolescence of its tools and systems. Moreover, standardisation bodies, like ISO, are inefficient and can not keep the pace with Internet evolution. As a consequence, Internet banking standards are primarily de facto standards. A last concern is the performance level of the web and the difficulties to predict its evolution. All these issues are causing serious problems to banks and make them unable to evaluate the economic model of Internet banking. Another danger for banks is the risk of desintermediation. But if this problem is considered as serious in the USA where important non-bank actors as Intuit are playing an important role, it has less impact in Europe.

In the Internet banking area, Europe is often considered as lagging behind the US but this conception is untrue. Two major reasons can explain that phenomenon. The first one is that American banks put principally the stress on retail electronic banking while European banks have these last years mainly been busy with corporate electronic banking. The second one is that an important effort of standardisation has been realised in the USA regarding Internet banking interfaces while the technical solutions in Europe are almost as many as the number of banks offering Internet banking.

However, according to new research by Bluesky International Marketing, when observing both markets, it appears that Western Europe has nearly twice as many transactional Internet banks as in the US. Germany, Spain, France and Holland are leading the on-line revolution in retail banking. Another fact to be stressed is that many of the European leaders in Internet banking are value-focused specialists and small, foreign and new players rather than the traditional banks.

Internet bank sites can be ranked into five categories: from simple or detailed "brochureware" to "interactive banks" which offer calculators and/or formatted e-mail communications,

Figure 1: European Internet Banks by Category (1998)

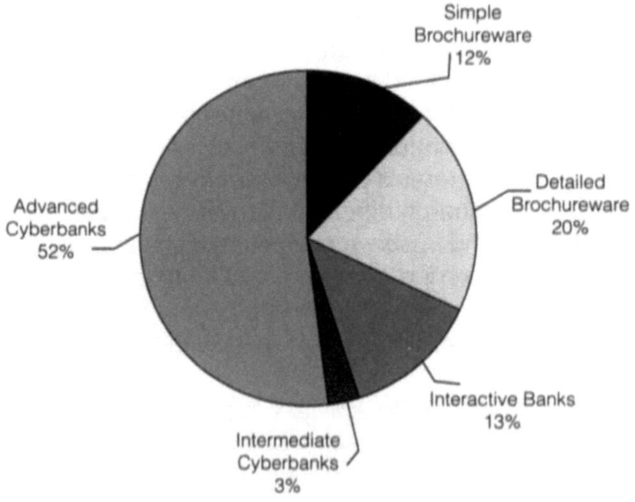

Source: Bluesky International, 1998

to "intermediate banks" where customers can consult their accounts and statements on-line, on to "advanced cyberbanks" which are transactional sites where customers can transfer money, pay bills and/or buy financial services.

According to Bluesky, there are 476 intermediate and advanced cyberbanks in Europe and only 249 in the US. Germany has the largest number of the 863 total sites. Although often considered as laggards in Internet usage, France and Spain combined have six times as many advanced cyberbanks as the UK, even though they have only half the number of internet users. Nearly all of the Internet bank sites in Holland are advanced cyberbanks, 40% of which are available in English as well as Dutch as they use the Internet to expand beyond their domestic borders.

Initiatives

Internet banking is developing and multiple initiative have been launched both in Europe and in the USA. In this section, major American initiatives will de detailed (Integrion and TransPoint) together with an European one, namely Argenvia.

Integrion

The first initiative is Integrion Financial Network, launched in September 1996 by fifteen leading North-American banks, including Banc One, Bank of America, First Chicago, Mellon Bank, NationsBank, Royal Bank of Canada, and IBM. Founding banks represent half the retail banking population in North America, over 60 million households.

IBM was the driving force behind the initiative whose key objective is to create a common network infrastructure, based on open standards, to provide a secure, high-performance, reliable transaction processing platform to enable electronic access between banks and their individual and small business customers. Integrion has moved quickly and impressively since its inception:

- In March 1997, Gold standard for Electronic Financial Services was published. Gold is the open, cross-platform messaging specification which initially support applications such as interactive banking and electronic bill payment.
- In August 1997, Integrion has acquired a subsidiary of Visa, Visa Interactive which processes banking transactions for about 60 financial institutions. The objective of the transaction was to enhance its capability to offer electronic bill payment services. As a result of the transaction, Visa has become a shareholder of Integrion.
- In September 1997, Citibank and First Union announced agreements to become equity owners of Integrion. As the result, share of U.S. consumer households covered by Integrion banks rose from 60% to 75%.

However, since the beginning of 1999, Integrion has lost some of its partners and shareholders. In April 1999, KeyCorp and Royal Bank of Canada were pulling out of Integrion. Currently, the network is governed by three banks. The three, which currently use Integrion technology platform, namely Bank America Corp, Bank

One Corp and Washington Mutual INC. Other owners who previously had a bigger interest, like Citibank, First Union Corp, Fleet Financial Group Inc, PNC Bank Corp and Wells Fargo and Co now would sit on a strategic advisory council.

TransPoint

The other major initiative came from Microsoft and it targeted the electronic bill payments, which is seen by many banks as a key component of online banking offerings, giving customers substantive benefits of cost and error reduction. Microsoft joined forces with First Data Corporation, leading third-party financial services processing company, and announced in June 1997 a joint venture, MSFDC, which planned to offer a fully integrated end-to-end electronic bill presentment and payment system, which link through Internet, the billers, the banks and the customers. On the 14 September 1998, when Citibank joined the venture the name was changed to TransPoint to better reflect the service it provides to consumers.

TransPoint targets a 50 percent to 75 percent reduction in sending bills and receiving payments in comparison to paper-based methods. In order to attract banks, Transpoint proposes an innovative business model. The service will be offered and branded by financial institutions as a part of their home banking service. Customers will not be charged for the service. Furthermore, unlike current electronic bill payment services that charge banks for payments made by their online customers, the TransPoint service includes no fees for the branding bank. Actually, TransPoint will share revenues with financial institutions that offer the service. There are also revenue opportunities for banks that offer the TransPoint service to their commercial customers. Billers will pay a small fee, comparable to the price of a postage stamp, for the delivery of an electronic bill and the return of remittance information and payment. TransPoint is operational today and rolling out the service gradually.

Argenvia

Argenvia is a Web retail banking service developed by Argentaria Bank (one of Spain's big three high street banks). Argentaria has more than 1 500 branches and holds over 29 million accounts for its customers. The web banking service currently support a range of facilities including payment of bills, transferring funds and sharing trades. Another feature uses push technology to alert customers when pre-set account thresholds are reached.
In July 1998, 7 000 on-line customers have already reached Argenvia. Argentaria expects that number to grow up to 40 000 by the year 2000. Customers are performing a whole range of electronic transactions with the Entrust/PKI (Public Key Infrastructure) providing highly secure authentication, encryption, verification and non-repudiation.

If Argenvia if one of the most advanced Internet banking systems in Europe, it must be stressed that a large majority a European banks are now providing their customer with more and more advanced Internet banking services. The success of internet banking services in many European countries is not to be proved anymore. In Belgium for example, BBL with its 13 000 connections a day has been recently forced to power-up its systems in order to keep pace with the increase of demand.

Standards

Even if Internet banking is still a fragmented market, some standards have been developed mainly on national scale but also on international scale.
This section gives a quick description of those standards together with an overview of their evolution. The first three systems in this section concerns American standards, namely Gold, OFX and IFX. We will also study the German HBCI system and the international SET standard from Visa, MasterCard and Europay.

Gold

The Gold Standard for Electronic Financial Services ("Gold") is the open, cross-platform messaging specification employed by Integrion Financial Network LLC for consumer and small business banking transactions within the Interactive Financial Services (IFS) platform.

Developed by the financial services industry, Gold and its associated programming environment support a wide variety of consumer access devices including Internet, intranet, personal financial management (PFM) software and telephone, enabling banks to offer customers a broad set of options through a standard interface.

Gold currently supports applications such as interactive banking and electronic bill payment, and will be extended to offer electronic bill presentment, securities and mutual fund transactions, and electronic cash. Conformance to the Gold Standard allows financial institutions and their developers to create custom applications at the client front-end or bank host while enjoying the benefits of a common message standard.

OFX

Open Financial Exchange is a unified specification for the electronic exchange data between financial institutions, business and customers via the internet. Created by CheckFree, Intuit and Microsoft in early 1997, Open Financial Exchange support a wide range of financial activities including consumer and small business banking; consumer and small business bill payment; bill presentment and investments, including stocks, bonds and mutual funds. Other financial services, including financial planning and insurance, will be added in the future and will be incorporated into the specification.

Open Financial Exchange, which supports transactional Web sites, thin clients and personal financial software, streamlines the process financial institutions need to connect to multiple customer interfaces, processors and system integrators. By making it more compelling for financial institutions to implement online financial services, Open Financial Exchange will help accelerate the adoption of on-line financial services by financial institutions and their customers.

The Open Financial Exchange specification is publicly available for implementation by any financial institution or vendor, and is being used by CheckFree, Intutuit and Microsoft as the primary mechanism for supporting financial data exchange in their products and services.

IFX

In early 1998, BITS (the Banking Industry Technology Secretariat) began to work on a new open framework for the exchange of financial data and instructions, converging the Integrion Financial Network's Gold and OFX specifications, called IFX. BITS is a division of The Bankers Roundtable, an organisation with a membership base of the largest 152 banks in the country, created to encourage greater choices in banking software, access devices, and the development of more efficient processing capabilities for the benefit of bank customers.

The converged specification will provide a robust and scaleable framework for the exchange of financial data and instructions independent of a particular network technology or computing platform. One of the BITS' major emphases has been to find ways to leverage the existing banking infrastructure. Overall infrastructure costs will be lowered: banks will be able to reduce costs by approximately 1 to 3 million USD (source: Integrion), because they will no longer have to build two systems or update two software programs.

BITS currently continues to make steady progress. However a usable, full-featured converged standard is still several product cycles away. Until the converged standard has emerged, Gold and OFX will remain in use. For example, Integrion already accept OFX versions of Quicken 99 and Money 99. This connectivity is possible through an OFX to Gold translation server that the companies have developed. This connection is an interim solution to allow for interoperability while the converged Gold and OFX standards is being developed.

HBCI

In 1997, German financial institutions have agreed upon a common homebanking standard, called "Homebanking Computer Interface (HBCI)". Launched in 1997, the new standard was developed to provide customers and developers of homebanking software a common interface for communication with financial institutions (multi-banking capability).
Above that HBCI increases the security of on-line banking. The new standard is based upon cryptographic security mechanisms. HBCI can be used via any on-line service and network, as well as via direct links with banks. Notable manufacturers of homebanking software have already announced to support HBCI in their products. The German financial sector perceives HBCI as the homebanking standard of the future.

Currently, HBCI continues to develop and German financial sector is convinced there is a lot of business cases provided by financial institutions to their customers, that are not yet standardised by HBCI (version 2.0.).

The following business cases are scheduled for the next version: Brokerage, foreign exchange rates, scheduled summary transfers/debits, order of cards/ credit card account overview, call for single sales, standardised international payments (non-SWIFT), submission of merchant sales by GeldKarte and file transfer.

SET

The SET payment standard has been developed by Visa, MasterCard and Europay. It is based on the digital certificates technology (Public Key Infrastructure). It aims to allow customers to pay on Internet using their credit card without sending their card number through the network.

Since April 1997, SET programs have been happening around the world in 39 countries. In Japan, Smart Commerce Japan implemented a pioneering electronic commerce program in stores and on-line. In Europe, the Bank of Ireland was first with a SET pilot, in the spring of 1997, followed closely by SET programs in Spain and Finland. Subsequently, SET programs were implemented in Austria, Sweden, Norway, Denmark, Italy, Germany, Portugal and France. In total, 47 Visa financial institutions in 18 countries are participating in European SET programs.

But SET pilots are still facing a series of technical problems. The protocol is complex, which adversely affects system performance. SET is built around a classic four part model: Issuer / Cardholder / Merchant / Acquirer. From the technical perspective, the performance is inadequate, with heavy system overhead and sluggish response time. The four parts are independent and can thus supplied by up to four different providers. Interoperability can thus become an issue. The protocol is constantly evolving, with version 2 expected soon. However, many vendors and financial institutions are already implementing new features (debit and chip), which may result in standard fragmentation. An example is the development of a SET system based on the smart-card technology (Cyber-Comm). This system aims to allow a much more efficient access to transaction security management.

Issues

Currently more and more advanced Internet banking systems are being developed by the private sector. However many questions are still open:

- Do the different system provide banks and their customers with an adapted security level?
- Does Europe adapt or adopt the American standards?
- Do the system refer to a card number or to a bank account?
- Are hardware based system more secure than software based ones?
- What is the dynamic of the co-opetition game on the Internet banking area (banks vs. non-banks and banks vs. Banks)?
- What is the real impact of Internet banking on the customer loyalty?
- How will the systems evolve towards end to end connectivity?

Those issues will be detailed in this section.

Security

The security risks associated with transactions over the Internet receive a lot of attention. At the International Banking and Information Security Conference 97, it was evident that banks, although very interested in Internet banking are also concerned with the risks connected with procedures for transactions over the Internet. Today, banks are already loosing enormous amounts through cheques and credit card fraud. The security solutions of the future are therefore high on the bank's agenda. If customers distrust the security it may create multiple problems. Banks will find it hard to launch Internet banking services if demand is low because of security doubts. Banks are in some senses trend setters in determining the technology that will be used in the future.

There are three key security aspects in a transaction: content confidentiality, integrity and authentication and non-repudiation. These aspects are treated independently, with various and often disparate standards. For Mr. Anderer (Brokat) this leads to fragmentation and inefficiency risk management. He recommends an integrated approach, providing security at all levels: client interface, transport and internal systems.

For client interface and transport, security is currently mainly ensured by the use of cryptographic instruments and by the set up of private financial network.
There are two kind of cryptography in use: the symmetric one (using the private key technology) and the asymmetric one (using the public key technology). If the first one is less secured than the second one, the disadvantage of the public key is that it requires more memory and can cause problems in the system management, like with SET. This explains why systems like TransPoint are still using SSL.

Private networks are another solution to secure transactions. Those networks are used both for corporate banking (ISABEL in Belgium) and for retail banking (Integrion in the USA). To provide its customers with increased security, Integrion has decided to develop the Interactive Financial Services (IFS) platform which is part of the IBM network computing platform. The IFS platform was developed by IBM based upon Integrion's requirements. It is operated by IBM, who is the exclusive provider of infrastructure and platform services to Integrion. The Integrion platform interfaces with the IBM Global Network, which already securely handles a large volume of electronic payment transactions. This network will also support the infrastructure for Integrion's IFNtm services, providing a secure and accurate delivery channel.

De facto standards

While a need for official internet banking standards is generally acknowledged, standard development process is moving too quickly for the official standardisation bodies such as ISO or CEN. Thus standards are developed either by technology providers or by financial institutions and their common organisations. As this situation is unlikely to change, the problems of co-ordination and interoperability will maybe become more serious. However, efforts developed by groups leaded by the private sector, like BITS in the USA, are encouraging compatibility of systems but only at the national level. An international co-operation may soon become highly desirable in order to harmonise the different national standards.

European position: Adapt or adopt
The challenge for Europe is to move from a mosaic of national standards to a set of common standards, which reconcile national differences and the need for global interoperability. This challenge is particularly critical at the same time when the European Union is facing a combined pressure of cyberfinance and eurofinance. Within few years, retail banking within Europe will become cross-border and Internet banking is a major vector of this trend. In addition to common standards, there is a need for common approach to the trust infrastructure, in order to avoid confusion of rules for digital certificates and certification authorities.

Internet banking is currently seen as primarily retail customer driven, even if national systems for corporate banking exist too (ISABEL in Belgium, for example). Thus internet banking standards are primarily national and reflect local banking environment. In Europe, common internet banking standards were developed for instance in Germany and in Spain. Similarly the two internet banking standards developed in the USA, OFX and Gold, are based on local banking environment. Many experts are concerned about international applicability of such "home-grown" standards.

However standards are evolving, and more and more features are added to them. For example, in the US, banking and brokerage are often considered as two distinct activities, while in Europe they are often integrated. Not so far ago, Gold and Integrion do not allow on-line brokerage operations. This is not true anymore. Since 1997, leading OFX banks, brokerage firms (including Charles Schwab) and application companies have collaborated to develop brokerage applications compatible with the OFX standard. In September 1998, OFX has integrated a new server technology and implementation services from Entreprise Engineering Inc.(EEI), enabling brokerages to deliver information and transaction services over the Internet via Quicken 99 software.

Even if it appears that Western Europe has nearly twice as many transactional Internet banks as in the US, no real European standardisation effort has been observed yet, while in the USA, the two major standards are currently converging towards a compatible solution. As we have seen it, even almost no national standards have been developed in Europe. And pressure of Eurofinance and Cyberfinance becomes stronger everyday. It hence seems obvious that European banking sector will soon become confronted to choices concerning the development of compatible systems. Those questions are:

- Must European banks develop a common European standard?
- On which (existing ?) system will it be based?
- Must European banks adopt the American standard?
- If yes, will they have to adapt them in order to make it match with European realities?

Co-opetition game
The threat from new competitors makes it all the more important for banks to use IT. The entry of new competitors and niche banks does without doubt pressure the

banking sector to change. If bank customers are willing to change their behaviour, traditional banks must continue their development of IT, competing or collaborating with non-banks new entrants.

An example of that phenomenon is the convergence of OFX and Gold. The impetus for OFX and Integrion was provided primarily by technology suppliers. But US financial institutions, through BITS, have become more actively involved in standards development and contributed to the convergence of the two standards. According to Catherine Allen (BITS), the convergence between Gold and OFX demonstrates clearly the new leadership of the banking industry working with technology players, in shaping the framework of electronic commerce. Another example proving that American banks are seriously beginning to take the reins in internet banking is the acquisition of Visa Interactive by Integrion. When buying a small home-banking and bill-payment business in 1994, Visa wanted to propose to its members a common systems for internet-banking, bearing the Visa brand. Visa designed a own personal standard computer software in order to allow Visa's smallest member banks to offer their customers the same services that Visa's biggest bank provided. But Visa had to give up the system, not for problem of performance of the technology but because of political struggles. Visa Interactive was not satisfying the strategic needs of the largest owners of Visa. Major banks refused Visa to use a part of their funding to give a level playing field to smallest banks. This is an example that if standardisation is an essential condition, product differentiation and comparative advantages remain crucial too.

As for other banking technological products, the banking sector is playing a co-opetition game in the internet banking sector. On one hand, banks and non-banks co-operates in order to create common infrastructures to propose their customers interoperable systems and to decrease their development costs. On the other hand, they compete very aggressively to differentiate their products in order to develop a comparative advantage regarding their competitors. This phenomenon equally takes place between banks.

A last example is the recent modifications in Intergrion's capital. Many banks are now partly or completely retiring of Integrion. According to International Data Corp, it is a strategic move on the part of banks not to get too involved in the future development of the consortium but maintain a working relationship with it. Some of the banks might prefer to develop electronic banking independently, or would rather not pay so much in a common system.

Customer Loyalty decrease

Internet banking shift the power to the customer. As a matter of fact, with emergence of the standards in the Internet banking area, it will become easier for the customer to compare services from different banks and to change for the better one. Development of innovative systems in the Internet banking area in order to attract customers has hence become a priority, even if those services are often not the most interesting for banks. It is true for example that the margin on payment services is low and does not encourage banks to innovate in that sector. However, payment services also constitute the most important link between banks and their customers. The banks need this in order to attract customers with higher margin. Those high margin services can be either innovative Internet banking services or services distributed by other channels.

Card or Bank account

Some internet banking systems are based on a card and not linked to any bank account. Such systems potentially reinforce the decrease of customers loyalty towards their respective banks. Because their card allows them to easily access banking services of all banks issuing and acquiring the card instead of being linked to a single bank account.

Concerning SET, many banks feel that they do not have an inadequate input into the decision-making process. While they recognise the usefulness of a card-based payment approach, such as SET, they believe that internet banking should emphasise direct relations between banks and their customers and thus give high priority to account-based payment mechanisms.

Furthermore, such direct relationship reduce the need for elaborate certificate and Public Key management arrangements, which are geared primarily for occasional purchases.

Software or hardware solutions

Concerning specific security solutions for banks transactions over the Internet, there is a strong support for smart cards solutions. According to a 1998 study of iD2 Technologies, a strong majority of banks views smart cards as the solution giving the highest level of security and reliability. A majority of the banks believe that smart cards based security solutions will be the most widely used solution in Internet banking in the future. However, some banks are still concerned about the pricing of existing smart card (compared to magnetic stripes ones). A significant reduction in prices may be necessary for the technology to win widespread acceptance.

Although the banks themselves believe that the security levels for bank transactions over the Internet are sufficient, they also believe that their customers distrust existing security solutions, primarily because they are software based. The big concern about this kind of product is to rely on a single software solution which only identifies the user's computer but not the cardholder or in a more general way the customer. This may cause serious problems if a third person obtains (by a way or another) the password to activate the certificate. For example, most of Internet users connect themselves on their working place, essentially for connection costs. It means that many persons have access to the computer in which the certificate is recorded.

Hardware based solutions using smart card can avoid those problems by locating the certificate on the card. Such systems allow card identification instead of computer identification. There are already two SET pilot standards implementing the chip card technology in France and in Japan.

In France, banks and GIE Cartes Bancaires agreed to combine their smart cards projects and to define a single secure solution called CyberCOMM. CyberCOMM is an Internet banking solution based on the existing French smart card-based debit card and using the SET security protocol. Such solution should ensure a world-wide interoperability. Its deployment is planned to begin mid-1999.

Smart Commerce Japan will also test the full scope of chip based payment systems in both the real and virtual worlds. The pilot is merging the specifications of the two technologies into one all-encompassing product using's Visa's recently released Chip electronic Commerce Standard. The Chip Electronic Commerce Standard is compliant with the EMV standard for global chip card interoperability and the SET Protocol.

Towards end to end connectivity
Another issue in the deployment of internet banking standards is the need for differentiation among various standards. User interface standards are different from back-office standards for connection between internal banking systems. Those various standards need to be interoperable and easy to integrate in an overall framework.

Even if systems like Integrion already offer end-to-end connectivity, they must be adapted each time they are liked to new front-end or back-end systems. On the front-end side, many efforts have already been made. Gold and OFX are both supporting widespread Money and Quicken software's. On the back-end side, no standard exist. There are as many systems as banks and adaptations are to be made.

A solution to that problem can be the new XML language. XML is a markup language for documents containing structured information. XML was created so that richly structured documents could be used over the networks. The only viable alternatives, HTML and SGML, are not practical for this purpose. HTML is not flexible enough while SGML is too difficult to implement for a web browser. This new language constitutes an opportunity for banks to use a richer language helping to bridge front-end and back-end in an accurate way. What is currently impossible with HTML language.

Next steps
Internet banking is taking an increasing importance in banks policy. Systems are being developed throughout the USA and Europe. However many issues are still to be discussed and better understood. For that reason we think that it would be useful to create an exchange forum in charge of the dissemination of information on Internet banking development within Europe. FIWG, through its Web site, could provide the necessary support for such an exchange.

Internet Banking systems in Europe are almost as numerous as the number of banking institutions. FIWG thinks that it would be interesting to set up a workgroup which would study the possibility of the development of a European standardised user interface for Internet banking. This group would concentrated on the technical implications of a common interface and also with the political implications of such a system (competition level in this area, national specificity, ...) .

Another interesting role of the workgroup would be to determine what extent US and European systems can converge to a common standard and under which conditions. The participation of actors such as BITS should hence be welcome.
It would also be interesting that the workgroup deals with security aspects of Internet banking and especially the opportunity of implementing the smart card as a secured key for Internet banking operations.

A last crucial point on which exchange would be welcome is the development of back-end systems interface. It would be interesting to determine precisely the specificity's of internal systems (amount of data exchanged, level of transaction security) in order to help European banking sector to elaborate common specifications concerning a potential standardised back-end interface.

To facilitate substantive discussion and progress, specific subgroups and/or taskforces should be set dealing with topics such:

- User interface: to what extent can various standards (US and European) converge to a common standard
- Security standards
- Internal systems interfaces

Internet Banking – An Overview

Title: *Internet Banking – An Overview*
Author: *Juergen Seitz and Eberhard Stickel*
Abstract:

Juergen Seitz is Assistant Professor of Information Systems at Viadrina European University Frankfurt (Oder). He studied economics and business administration at Hohenheim University, Germany. His main research interest include IT use in banking and electronic payment systems.
E-mail: jse@euv-frankfurt-o.de
Eberhard Stickel is Professor of Information Systems at Viadrina European University Frankfurt (Oder). He studied mathematics and computer science at the University of Ulm, Germany and at Syracuse University NY, USA. He is holding a Phd of Ulm University. His main research interests are economics of IT use, virtual banking and electronic payment systems.

Introduction

Consumer behavior is changing partly because of more spare time. The way of use of financial services is characterized by individuality, mobility, independenceof place and time, and flexibility. Financial transactions caused by purchases will more and more be carried out by non- and near-banks. These facts represent big challenges for providers of financial services. More and more the Internet is considered to be a "strategic weapon".

Financial services companies are using the Internet as a new distribution channel. The goals are:

- complex products may be offered in an equivalent quality with lower costs to more potentialcustomers;
- there may be contacts from each place of earth at any time of day or night.

This means that financial institutions may enlarge their market area without building new offices or field services, respectively. Because of its image as an innovative corporation, better interacting possibilities, the usage of rationalization potentials, promotion of self-service ideas, the improvement of its competitive situation by development of core competencies together with the construction of market entry barriers, it may be possible to increase profits and marketshares.

One way of exploiting rationalization potentials is the implementation of the entire transaction (from purchase to payment) under a common user interface. Information collected in operative databases of financial institutions allows them to act as information brokers. Offering special information in closed user groups may result in more intense customer commitment,as well as customer bonding. Know-how that is built up by Internet presence may be used to facilitate Internet presence of smaller companies. The use of digital coin-based money to completely settle transactions in the Internet is a new service provided by financial institutions.

The presentation is organized as follows. In the next section the properties of the Internet as a distribution channel are explored. In section 3 it is discussed how financial institutions currently use the Internet. It is shown that the possibility to carry out safe transactions is crucial.The focus of the presentation in the fourth chapter lies on the discussion of electronic payment systems. They may be seen as a prerequisite for more complex safe transactions.

Internet as a Distribution Channel

Distribution channels are physical capacities to build up customer contacts in a systematic way in order to inform, counsel and sell products and services [Aus96]. Like America Online or Compuserve the Internet is a so-called electronic distribution channel. Combined with self-service terminals and telecommunication equipment electronic distribution channels are technical channels within the class of media distribution channels. Another example for a media distribution channel is direct mail.

Figure 1: Systematic of distribution channels [Aus96]

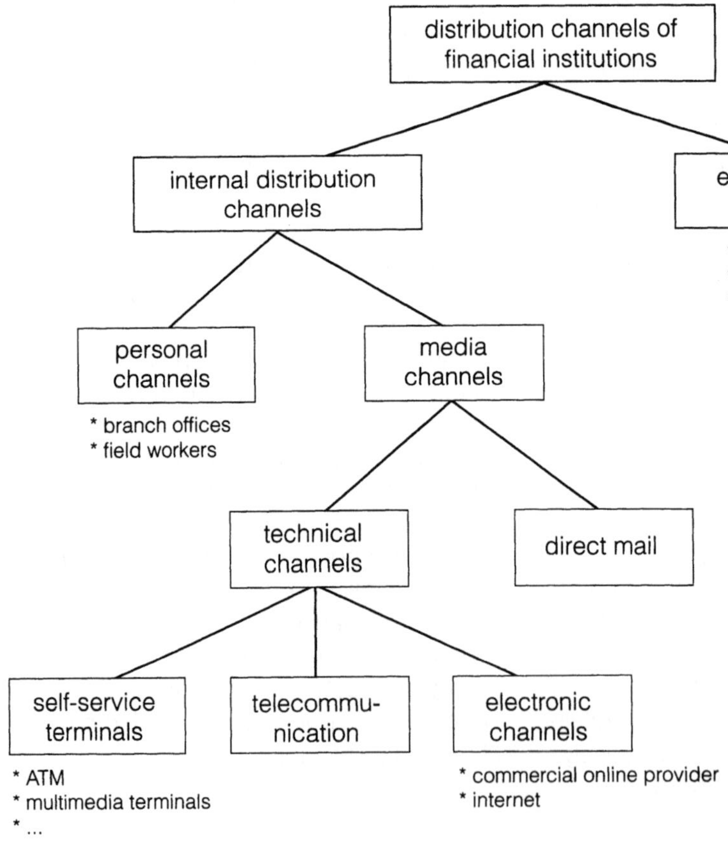

Today, media distribution channels are an important way of distributing information and managing standard transactions. Counseling is mostly done in branch offices or by field workers. Together, personal and media distribution channels are called internal distribution channels. On the other side there are external distribution channels like salesman or franchising partners. Figure 1 visualizes this classification.

The worldwide web (WWW, 3W,W3) is the most wellknown and most important Internet service. A standard user interface to be able to address a large number of users was one of the development goals of the WWW. The WWW is a worldwide network consisting of a large number of various computers. The user interface integrates other Internet services like ftp, telnet, email, etc. The WWW is based on hypertext and hypermedia principles. Therefore, it is possible to present information in a well structured manner. Documents are connected via links. Besides text documents it is possible to include pictures, sound and videos.

The client/server-architecture forms the basic implementation platform of the WWW. Data are stored on a WWW-server. The server software responds to inquiries from WWW-clients and sends files to the clients. The files may be static on the server or build up dynamically by means of parameters.The client interprets the files and presents the information on its screen. In modern browsers features allowing execution of application modules on client computers are implemented. Corresponding programming languages are e.g. Java, JavaScript or ActiveX.

Communication between client and server is done through the hypertext transfer protocol (HTTP). HTTP is a very simple protocol. It allows short(er) response times and reduced use of the server. On the other hand, a connection has to be build up for each inquiry. Each document is addressed by a unique key. The uniform resource locator (URL) is formed by the address of the server, the (directory) path and the filename. Sometimes it is useful to append further data, e.g. to control programs that may be executed.

Areas of Use of the InternetIn Financial Institutions

Generally we may distinguish four classes of Internet use in financial institutions:

- information presentation
- information presentation together with two way (asynchronous) communication (e.g. email to request further information)
- interaction with user (e.g. execution of programs with individual customer data)
- transaction banking (e.g. electronic payments).

Information Presentation
Information may be provided in connection with one or two way communication. One way communication means that the institution uses the Internet only as a presentation medium for its products and services. The simplest way to use two way communication is to allow users to send electronic mails to the server in order to ask for further information or make suggestions with respect to the Internet site.

Interaction with customers requires quick information exchange. Information provided by the user controls the information offered by the server. If the customer is identified and authenticated connecting to operative systems of the financial institution may be possible. Then, often very little information has to be provided by the customer since datastored in the databases of the financial institution may be used.

Presentation of product information may be used to initiate new contacts. Implemented product models permit the construction of optimal insurance or financing contracts by using simpler components [Sei97]. Using mathematical models the customer may analyze his portfolios. To do so, he may use simulation techniques, what-if-analysis and other similar techniques.

Most Internet presentations by financial institutions fall into one of these three categories (actually most of them are within the first two groups). If actual contracting is desired transaction management is necessary.

Internet Transactions
There are a large number of different financial transactions, like e.g. customer payments, securities transactions applications for loans or insurance acquisitions.

Due to the structure and the intention of the Internet to be an open network high security risks are involved with financial transactions. Today, various techniques and standards are offered in order to control or even avoid these risks. Basic requirements are as follows:

- Customer and financial institution have to authenticate each other.
- Private data have to be encoded. Cryptographic algorithms used need to have certain characteristics. No third party should be able to quickly get access to messages or even to divert financial transactions.
- A digital signature is necessary to get binding legal contracts. These digital signatures have to secure the integrity of signed documents. It needs to be guaranteed that sender and receiver have the same intentions.

Based on these requirements HTTP is extended to S-HTTP. Because the security level of this protocol still is not high enough various additional techniques and standards have been developed. Examples are the home-banking communication interface (HBCI), secure electronic transactions(SET) or secure socket layer (SSL).

Different types of methods are used or currently tested. These methods may be classified into hardware- and/or software-based solutions. Hardware-based solutions use a chip that is physically located between computer and keyboard. Such a chip is unique. Note, that the system is hardly usable with laptop computers and/or with different financial institutions.

More often software-based methods are used. Personal identification numbers (PIN) identify the user. For each transaction a transaction number (TAN) is necessary. Data are encoded using algorithms like

IDEA with a 128 bit-keyor RSA with a 1.024 bit-key. A higher level of security maybe reached by means of a so-called electronic fingerprint. This fingerprint is taken before and after the transmission. Then, both versions are compared. In case of any differences the transactionis aborted.

Digital coin-based money

Basic Requirements
Digital money may only be used for electronic commerce in an efficient and effective way if an infrastructure on a high technical level exists. A large transmission rate is a prerequisite for simultaneous transmission of product information to potential customers. Access has to be simple and economical. Therefore, private households need PCs with suitable software. If these technical conditions are met security problems have to be addressed. An excellent survey is the book by O'Mahony et al. [OMa97].

Actors Involved
New payment systems like digital coin-based money are only successful if they are accepted by a large number of persons. To get this acceptance all actors involved should have sufficient benefits that exceed their costs.

- The primary interest of customers is to carry out purchases in a comfortable manner. Associated payments should be possible from home in a simple and efficient way.
- The dealer usually has to bear the costs of payment transactions. On the other hand he may benefit through an improved image (being innovative) and possibly higher sales. Also, he may be able to reduce branch offices and sales personnel.
- System architects are responsible for the development of payment systems. Their benefits consist of royalties and service fees. Wide acceptance and usage of their system is an important requirement.
- System providers are intermediaries. Dealer's sales are forwarded to the financial institution. The system provider is responsible for transaction clearance. Moreover,he provides supporting services (problem management, user training). Again, revenues consist of fees and service charges.
- Financial institutions may promote certain systems. As was already mentioned, confidence in electronic payment systems is a key success factor. Note, however, that a large number of financial institutions, typically smaller ones in regional markets, may have difficulties to participate due to high costs and know-how required.
- Trust center control digital signature keys. They are responsible for the integrity of transmitted data and the authenticity of contractors. Trust centers help to secure confidence in a certain payment system.Their revenues consist of royalties and other service fees.

System Requirements
The following requirements are implementation independent. They are useful in comparing different payment systems.

- Open systems need security features to manage electronic payments. Security may be realized by cryptographic methods in connection with transaction numbers (TAN).
- A large number of customershave to be able to simultaneously carry out payment transactions.The systems have to work with a large number of customers and should be easily expandable. Hence, scalability is an important criteria.
- Small and smallest payments (micro- and pico-payments) should be possible. The corresponding accounting systems have to be efficient and effective. Therefore,it is necessary that costs per payment are low.
- The system has to be transparent. Customers have to be aware that payment actually takes place. System usage has to be simple.
- Private households should be able to accept digital money (micro-merchants).
- Electronic data transfers,as well as electronic payments may be traceable. Then it is possible to analyze e.g. payment information and to construct detailed customer profiles. In most cases customers wish to stay anonymous.
- Digital coins consist of a number of bits. Hence, there is some possibility that copies of the coins are made and eventually put into circulation. This phenomenon is known as the double spending problem. Hence,the payment system needs mechanisms to recognize and/or prevent repeated payments with the same digital coin.
- Digital coin-based payment systems have a nominal value problem. In order that a certain amount may be paid either coins with suitable values are needed or the payment system has to generate change in the form of new coins. Alternatively, all coins have the same smallest possible face value (e.g. one cent or even a fraction in the US). The last alternative requires that a large number of digital coins have to circulate and to be checked for authenticity.
- Digital money should be convertible into "real" money, whenever this is desired.
- Confidence in an electronic currency means that stable exchange rates between electronic and"real" currencies are necessary. If the exchange rates are unstable there is a chance of arbitrage profits. This, in turn, would reduce confidence since losses due to value fluctuations are possible.
- Digital money is stored locally on hard disks or other media. In case of disk crashes or other problems mechanisms to recover the original state are necessary.

Basic Principles of Electronic Coin-based Payment Systems

Due to the increasing importance of electronic commerce via the Internet the importance of digital money increases. Representing "real" money in an electronic world means that properties and functionalities like anonymity,authenticity, as well as availability of pico-payments are considered. Like "real" money, digital coins have an inherent value.

Depending on the way digital money is implemented there exist different cryptographic methods and organizational precautions to avoid the usage of forged money. Basically, there are two different types of digital coin-based money:

- Using specific cryptographic method the anonymity of digital money may be achieved. Then, neither the financial institution nor the dealer may build up a connection between the customer and coins used by him. The financial institution only knows to which customer the coins are transferred initially [Cha92, p. 96].
- Coins with customer identifying characteristics allow the financial institution to identify the customer and to follow up on payments where the coin has been used in.

Also, the payment process may be classified into online and offline transactions. Figure2 summarizes the different approaches.

- If an online payment takes place the coins will be checked immediately for authenticity.This implies that a digital coin is used only once. The financial institution needs to check the authenticity by using a list of all coins that have been issued or a list of all coins that have been sent in for credit.
- In case of offline payments the coins may be used more than once. To avoid double spending it is necessary to store information about the user or the users on the coin in order to be able to perform checks later. Anonymity may be guaranteed by so-called secret sharing. Then, the financial institution only gets information in case of double spending.

Examples of Digital Coin-based Money

ECash

ECash is anonymous digital money whose validity is checked online by the corresponding financial institution. ECash is developed by DigiCash and is offered by Mark Twain Bank, St. Louis since 1995. Deutsche Bank AG, Frankfurt (Main) offers eCash as a pilot project to its customers since October 1997.

The customer withdraws digital money from his eCash-account using the so-called blinding method and stores it on its harddisk. The blinding method works as follows.The client encodes a serial number and sends it to the financial institution. The financial institution certifies the coin and transfers it back to the customer. The customer then decodes the serial number. Hence, the serial number is not known to the financial institution, which guarantees anonymity. In order to avoid double spending the financial institution has to record the serial numbers of all incoming coins. At each purchase via the Internet the customer gives digital coins to the dealer.

Figure 2: Security approaches to avoid copying digital coins [Sch97, p. 59]

	Offline payments	**Online payments**
Anonymous digital coins	Secret sharing by storing some information on the coin	Blinding and immediate check by the financial institution
Coins with identifying characteristics	Storage of information about the transaction on the coin	Immediate check by the financial institution

The dealer immediately transfers the coins to his bank in order to check for validity. The dealer'sbank registers the numbers of the coins issued without tracing them back to the customer. Finally, the dealer is credited and delivers products and services ordered [Pan96]. Figure3 shows the payment process.

Digital coins may be used only once. ECash may be considered to be a currency of its own. Financial institutions have to use special accounts. They also guarantee conversion into "real" money. As a consequence central banks like the Bundesbank or the Federal Reserve Bankhave difficulties in controlling money supply (financial institutions may create additional money and thereby increase the amount ofmoney supplied; this is well-known in the case of so-called checkbook or deposit money [Eva92,&bnsp;p. 410])

ECash security is achieved by using an asymmetric cryptographic algorithm. Account access may be protected additionally by using personal passwords. The storage of a coin's serial numbers does prevent double spending.There may be a problem with scalability, however. The costs of checking for authenticity of coins are relatively high because the check have to be done online. This means that the suitability for micro- and pico-payments has to be evaluated carefully. Each person who has an eCash-account may accept eCash coins. The blinding method, as was already indicated, guarantees anonymity.

NetCash

The NetCash method is developedat the University of Southern California. One important goal of this project is the use of already existing accounting systems and procedures in financial institutions. This reduces initial investment costs. In contrast to eCash, this method is based on a decentralized approach. Consequently, problems associated witha large number of coins and participants may be solved more

Figure 3: Payment process with eCash

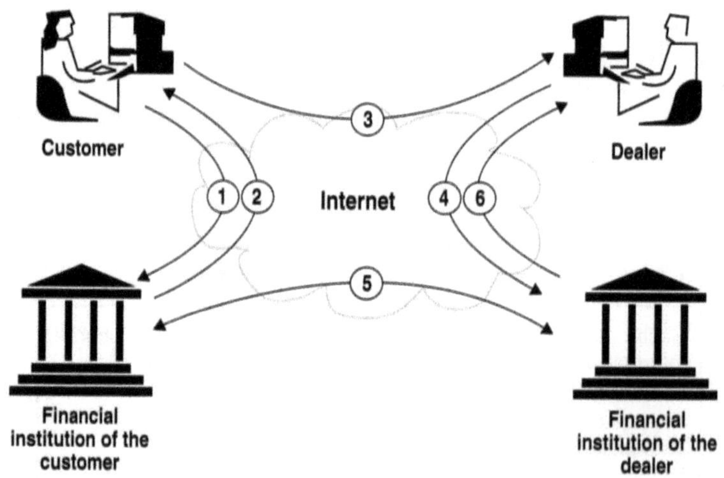

1. Customer sends coin with encoded serial number to the financial institution
2. Financial institution sends coin back signed
3. Customer decodes serial number and uses the coin for payment
4. Dealer sends coin to his financial institution
5. Check and clearing
6. Dealer gets confirmation and is credited

easily.Therefore, reduced anonymity is accepted and the cooperation of all participating financial institutions is required.

The system is based on independent distributed currency servers. Currency servers are locations to exchange anonymous into non-anonymous money. Each currency server possesses an account on an accounting server. Clearing is done by the currency server. It is necessary that the integrity of the servers is certified and that currency servers accept coins from other currency servers. NetCash-coins have a face value and a serial number. Also, the address of the issuing server and an expiry date is stored.

Figure 4 shows the payment process using NetCash. The customer gets NetCash-coins from a currency server. These coins are encoded with a public key and sent to the dealer. Anonymity of the customer may be guaranteed by using a new session key for each message. The dealer transfers the coins received immediately to his currency server. From the currency server he either receives new coins or the corresponding value will be credited to his account. Final clearing is done by the currency server. The serial numbers of all coins that are not sent back and are not yet expired are stored on the currency server in order to avoid double spending. This means reduced anonymity. Anonymity may be increased by exchanging the coins at another server. Security is reached by means of a hybrid cryptographic algorithm. Like eCash we have a method that requires a lot of communication. The usage for micro-payments, however, should be more efficient. Each person may accept NetCash-coins because the system allows free exchange of coins.

Millicent

The Millicent method is developed by Digital Equipment Corporation (DEC) to manage small and smallest payments (e.g.

Figure 4: Payment process using NetCash

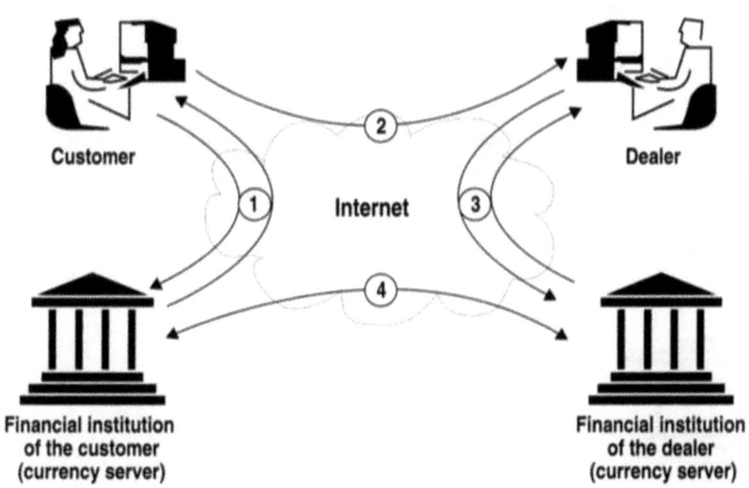

1. Customer gets coin by payment (e.g. by credit card, check, ...)
2. Customer sends coins encoded in a asymmetric way to the dealer
3. Dealer presents coins and gets new coins or is credited; dealer sends confirmation to the customer
4. Clearing by financial institution

payment for getting information from the Internet about news and stock quotations or payment for small programs like Java-applets).

The customer buys a brokerscrip with a defined value by using his credit card or by debiting a suitable bank or broker account. Such a scrip is like a telephone card. At the time of purchase the customer exchanges parts of the scrip into a dealer's scrip. This scrip is then sent to the dealer. The dealer collects all scrips and exchanges them into "real" money. Figure 5 shows the payment process using Millicent.

To guarantee the security of this method one-way-hash-functions that may be evaluated quickly (e.g. MD-5) are used. Furthermore, the costs of illegally decoding a scrip (this means finding the inverse of the hash-function used) are much higher than the scrip's value. A large number of transactions are possible at low costs compared to the other two methods discussed. In principle, each person may be registered at a broker and may then accept digital payments. There is no anonymity but there is the possibility to buy scrips from different brokers. Then, no comprehensive user profile may be built.

Conclusion

Currently most financial institutions use the Internet as a presentation medium. Often there is a possibility to request additional information or to perform individual calculations. Business transactions are rather rare at least in most European countries. On the other hand, a lot of effort is devoted to construct solutions to manage financial routine transactions like money transfers, opening and closing of accounts, implementation and deletion of standing orders and much more. Payment systems are developed to facilitate electronic commerce. In order to realize significant rationalization potentials no isolated but integratedsolutions that support existing business processes are required [Sei97]. Collaboration between competing financial institutions may be necessary to cut down development costs [Sti97b].

Figure 5: Payment process using Millicent

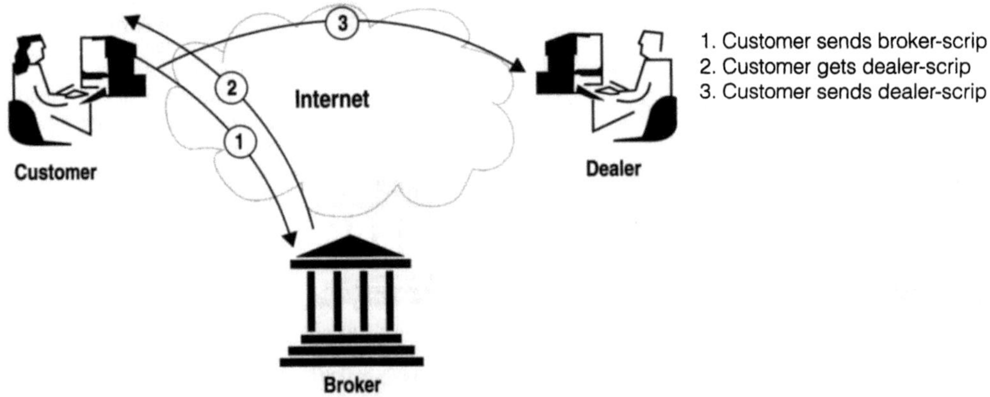

In general, financial institutions have to decide on their Internet presence. Is it worth to invest significant sums? It can be shown that there are not necessarily first mover advantages [Sti97a]. On the other hand, fast reactions to actions of competitors are difficult since significant know-how is required to quickly build up an Internet presence. This implies that waiting too long may be extremely harmful and expensive. Consequently, a good strategy should be to build up know-how by means of small or medium pilot projects. Actions of competitors, as well as the development of the Internet should be monitored closely.

References

[Aus96] Ausfelder, R.(1996): Die Einführung von Telebanking als Vertriebswege-Entscheidung von Kreditinstituten. Dissertation. Universität Passau. Frankfurt (Main).

[Cha92] Chaum, D. (1992): Achieving Electronic Privacy. In: Scientific American, August,pp. 96 - 101.

[Eva92] Evans, J. (1992): International Finance. A Markets Approach. Orlando.

[OMa97] O'Mahony, D., Peirce,M., Tewari, H. (1997): Electronic Payment Systems. Boston, London.

[Pan96] Panurach, P. (1996): Money in Electronic Commerce: Digital Cash, Electronic Funds Transfer, and Ecash. In: Communications of the ACM, Vol. 39, No. 6, pp.45 - 50.

[Sch97] Schuster, R.; Färber,J.; Eberl, M. (1997): Digital Cash: Zahlungssysteme im Internet. Berlin u. a.

[Sei97] Seitz, J., Stickel,E. (1997): Cooperative Software Development Supporting Financial Product Representation and Distribution in the World-wide Web. In: Wojtkowski, G. & Wojtkowski, W. & Wrycza, S. (Eds.): Proceedings of the Sixth International Conference on Information Systems Development (ISD '97), Boise: Plenum Press, to appear.

[Sti97a] Stickel, E. (1997): Problems Associated with Cost/Benefit Analysis for Strategic Information Systems. In: In: Wojtkowski, G. & Wojtkowski, W. & Wrycza,S. (Eds.): Proceedings of the Sixth International Conference on Information Systems Development (ISD '97), Boise: Plenum Press, to appear.

[Sti97b] Stickel, E. (1997): Collaboration among Competitors - A First Economic Analysis.

Chapter 2: Electronic Banking in Practice

Electronic Banking Developments

Title: *Electronic Banking Developments*
Author: *Richard Raysman and Peter Brown*
Abstract: *This article highlights some of the technical and legal issues raised by the Internet's expansion into banking.*

ELECTRONIC BANKING DEVELOPMENTS

The phrase, "cash, check, or charge" still has a comforting ring in the ears of merchants, consumers, and bankers alike. These common payment terms are readily understood by anyone who has gone shopping or deposited money into a checking account. Lawyers have mastered all of the rules surrounding these terms and feel comfortable counseling their banking clients.

But traditional banking terms have recently been modified by a few modern adjectives. Cash has become "CyberCash®". Check has become "electronic check." Banking is now "online banking." Signature verification has become "digital signature verification." Suddenly, the electronic banking revolution has made bank cards act "smart" -- even as it leaves bank executives and their counsel feeling "numb."

The Internet has the potential to change the very nature of banking and the growth potential for online consumer banking is enormous. According to some estimates, there are over 500 banks operating on the Internet in the United States, performing a variety of banking services. By the year 2000, an estimated 7,500 banks will have operations on the World Wide Web. Developments in electronic banking are likely to have an impact on all banks. Many bankers believe that a large part of banking in the 21st Century will be conducted online.

The development of electronic banking has raised both technical and regulatory concerns. For example, increased interest in online payment has pushed the development of online payment systems. At the same time, financial institutions have been hesitant to implement these new systems until uniform industry and regulatory standards are in place. Online transactions also create certain security risks, and an increased potential for fraud. A bank's liability for Internet fraud is far from certain.

The growth of Internet banking has also raised a host of new legal issues. Moving from a cash, check, or charge economy to a world of Internet transactions will require the next generation of bankers and regulators to address issues including security, authentication, consumer protection and privacy. This article highlights some of the technical and legal issues raised by the Internet's expansion into banking.

ONLINE PAYMENT TECHNOLOGIES

Credit cards are the predominant payment mechanism for most online retail transactions. Using credit cards online, however, can be risky because the Web is an unsecure medium. As credit card numbers and expiration dates are routed from computer to computer around the Internet, the risk of card security breaches increases.

In response to consumer demand for online shopping services, MasterCard, VISA, GTE, and a number of information technology partners introduced the Secured Electronic Transaction protocol ("SET"), which provides a mechanism for securing credit card transactions over open networks (such as the Internet). Use of the new SET technology enables bankcard purchases to take place over the Internet with much the same ease and security encountered in ordinary retail transactions. SET uses cryptography to provide confidentiality of information, ensure payment integrity, and facilitate authentication of merchants and cardholders.

Several companies have developed new online payment technologies. For example, a company called CyberCash, Inc., located in Reston, Virginia has created a system that functions as an online transaction manager for buyers and merchants. The CyberCash® system permits secure online credit card, cash and debit transactions. Invoicing and credit card information is forwarded, in encrypted form, from customer to merchant to Cybercash.® CyberCash® then routes completed transactional information to the banking system for payment approval. In this way, all transactions and electronic receipts flow through CyberCash's® "digital wallet" (also known in Federal Reserve circles as "digital purse" technology"). This hub-and-spoke routing system, with CyberCash® at the middle and the merchant, the customer, and the bank at the periphery, helps to protect all parties against fraudulent transactions.

Another approach to electronic payment is the "e-cash" system operated by DigiCash, Inc., a Netherlands-based company. Working directly with banks, DigiCash allows potential online purchasers to direct their banks to send an encrypted e-mail representing predetermined denominations of cash. Instead of paying by using a credit or debit card, the customer uses DigiCash's fully transferable electronic funds to make his or her online purchase. The bank then credits the merchant's account upon redemption of the DigiCash payment.

Yet another electronic payment service provider, First Virtual Holdings, Inc., located in San Diego, California, handles customer financial transactions offline, using a private network linked directly to banks. Online customers open an account and are given a personal identification number ("PIN"), which they can e-mail to a merchant; the merchant then verifies the PIN with First Virtual. Before debiting the customer's account, First Virtual sends an e-mail confirmation to the customer.

One technological area in which banks have taken the lead is in the development of electronic checking. The Financial Services Technology Consortium's ("FSTC's") Electronic Check initiative is intended to deal with the rapid growth of the worldwide electronic marketplace. As a technical matter, the Electronic Check is essentially an ASCII text block with a cryptographic signature. Its particular value is that even in the absence of a pre-existing relationship between checkwriter and merchant, it allows for the payment of goods and services over an open electronic network. The Electronic Check will serve as an alternative to traditional in-person cash, checking and credit card payment schemes. In replacing paper checks with Electronic Checkbooks, banks and bank customers will still be able to use existing reconciliation and settlement systems for clearing check transactions. As with paper checks, the Electronic Check will be a discrete "information object" having all needed information for making a complete payment.

STORED-VALUE CARDS

Another electronic payment mechanism, the "stored-value" card, is already in use. Stored-value cards resemble credit or debit cards. They are typically plastic cards which contain a magnetic strip or computer chips that are capable of storing a variety of information, including digitized data representing money or value.

Stored-value cards differ from debit or credit cards in that, at the time of purchase, the cards do not deduct money from a checking account or add charges to a credit account. The cards are pre-loaded with the electronic equivalent of money (and, in some cases, may be reloaded from a bank account). In order to accept stored-value card transactions, merchants need special point of sale terminals able to verify the card and the value contained on the card, deduct the purchase price from the value of the card, and transmit the information from the merchant's terminal to a financial institution's computer network.

The simplest form of stored-value card contains a limited amount of money stored on a card, either on a magnetic strip (a "dumb" card) or on a microchip embedded in the card (a "smart" card). Cards utilizing magnetic strips can be discarded or re-loaded, once the value has been used up.Smart cards are capable of processing a wider variety of information.

The Federal Reserve is in the process of finalizing rules addressing whether stored value cards are covered by the Electronic Funds Transfer Act ("EFTA") 15 U.S.C. § 1693 et seq. In May 1996, the Federal Reserve Board of Governors issued proposed amendments to clarify the applicability of Regulation E (12 C.F.R. Part 205), which implements the EFTA, to stored value cards. 61 F.R. 19696. Among other things, the EFTA and Regulation E govern restrictions on issuance of debit cards and account access devices, disclosure of terms and conditions of electronic funds transfer services, documentation and accounting of electronic fund transfers, notice requirements for changes in terms and conditions, error resolution procedures and limits on consumer liability for unauthorized transfers.

The proposed amendments to Regulation E would explicitly exempt the majority of stored-value cards from most Regulation E requirements. Loading of value onto smart cards from a bank account would be subject to Regulation E. The central issue for the Federal Reserve board is whether a stored-value transaction involves an electronic fund transfer from a consumer's asset account (a transfer of funds, initiated by electronic means, that results in a debit or credit to an account).

The proposed rules distinguish between four types of stored-value card systems. Stored value cards valued under $100 would be exempt from all Regulation E requirements. "Off-line accountable" systems, in which the balance of funds is recorded both on the card and in a central data facility, would be exempt from most of Regulation E's requirements, but would be governed by the rules covering initial disclosures. "Off-line unaccountable" systems, in which a record of the value of the card is contained only on the card itself would be exempt from Regulation E's requirements.

One category of stored-value systems, "online" stored-value systems, would be subject to certain requirements. Online systems operate like traditional debit cards, in which a transaction is authorized by online communication with a central data facility. Such systems would not be subject to periodic statement requirements (provided account balances and transaction summaries are provided on request), error resolution notice requirements or change-in-terms notices. Online systems would be subject to the provisions of Regulation E governing unsolicited issuance of account access devices, limitation of member liability for unauthorized transactions, and initial disclosure requirements.

The issuance of a final rule with respect to stored value cards remains on hold, pending further study of the issue by the Federal Reserve.

SECURITY AND AUTHENTICATION: ENCRYPTION

Most of the online payment technologies now in use or in development rely, to some extent, on encryption technology to provide security and authentication. Encryption has been used to make online commercial transactions secure, authenticate electronic communications by providing users with unique digital signatures, and maintain confidentiality in online transmissions.

Encryption provides a method for communicating securely over insecure communications channels, such as the Internet, by encoding messages so that they can be read only by parties having the proper key or decoder. Encryption basically involves the transformation of a readable message called "plaintext" into an unreadable "ciphertext" via a computer program which translates the message using an equation or algorithm. Public key encryption, also called "asymmetric" encryption, operates by means of a pair of complementary cryptographic keys (one public, and one secret). Each key unlocks the code made by the other and either can be used to encrypt a communication. Decryption, however, requires the complementary key. In this manner, a party can use a recipient's public key to encrypt a message which can only be decrypted by use of the recipient's secret key -- thus ensuring that only the recipient could read the communication and eliminating the need for a secure communications channel.

Federal policy regarding the export of encryption software, or software capable of encoding data, is one of the most rapidly developing issues of the law of the Internet, and will be of concern to bankers as an increasing number of international transactions are completed electronically. In the past, the export of cryptographic products was regulated by the U.S. Department of State, under the authority of the International Traffic in Arms Regulations ("ITAR"), which defined cryptographic devices, including software, as munitions. 22 U.S.C. § 2778(a)(1). The ITAR prohibited export of encryption software containing key lengths greater than 40 bits.

In November 1996, an Executive Order (Executive Order 13026) was signed which announced that the Department of Commerce, rather than the Department of State, would have jurisdiction over the export of these technologies. Under certain circumstances, the new Regulations permit issuance of licenses for the export of encryption products containing up to 56-bit key length Digital Encryption Standard. An Interim Rule implementing the Executive Order became effective on December 30, 1996. 61 F.R. 68572. Several pieces of legislation relating to encryption are currently pending in the U.S. Congress.

DIGITAL SIGNATURES

Digital signatures employ encryption technology to authenticate electronic transmissions and may be implemented in Internet banking. In creating a unique digital signature, a bank customer must proceed through the following steps. First, the customer is given a device for generating a "key" (e.g., on a stored-value card). Second, he or she makes a "key pair," each part of which bears a mathematical relationship to the other. Then the customer uses his or her "private" key to encrypt and sign his or her electronic message, and distributes his or her "public" key to all potential message recipients. Privately encrypted and signed messages are then verifiable by anyone with access to the "public key." At the same time, the holder of the private key can be assured that online transactions made in his or her name, but without his or her digital signature, will not be validated.

The use of digital signatures in an online contract presents added legal complications because electronically signed documents do not generally satisfy most state statutes of frauds. Subsequent production of a printed document, signed manually by each party in the traditional manner, does little to facilitate or encourage electronic commerce.

A potential legislative solution to these problems of electronic commerce is to give the same legal effect and imprimatur to electronic signatures that signatures on paper enjoy. Utah became the first jurisdiction to enact a digital signature statute in 1995. Utah Code Ann. § 46-3-101 et seq. 1995), as amended by S.B. 188, 52d Leg., Gen. Sess. (Utah 1996). Since then, other states have followed suit, including California, Washington, and Florida. More recently, New York, Georgia, Illinois, and Massachusetts, have considered digital signature statutes. Still other states, including Connecticut, Delaware, Hawaii, Iowa, Louisiana, Minnesota, New Mexico and Wyoming, have enacted laws which deal in one way or another with electronic signatures. While the Utah and Washington state statutes are comprehensive, other states, such as California, have limited the use of electronic signatures to communications with governmental entities. Cal. Gov. Code § 16.5 (1996). Future state efforts to legislate in this area are likely to follow either the comprehensive Utah model, or the short statute model used by Florida, which essentially empowers the Florida Secretary of State to study the issue and develop recommendations.

One important policy issue covered by the Utah statute, which will need to be addressed by other states, is the role of licensed electronic signature certification authorities. Coupled with digital signatures, certification offers the benefit of both non-repudiation and verification of sender identity. In Utah, a subscriber who accepts a certificate from a licensed authority is deemed to have indemnified the authority for losses caused by subscriber misrepresentation, fraud, or the failure to disclose a material fact.

In the context of messages transmitted via an unsecure communications medium like the Internet, digital signatures, coupled with the use of a third party certification authority, allow message recipients to allocate the risk of error or fraud. Meanwhile, at the federal level, banks themselves have moved to pressure Congress to enact legislation allowing use of digital signatures and other electronic authentication devices for inter-bank transactions. The driving force for the banks is not just that digital signatures offer a convenient way to authenticate and verify electronic documents, but also that the enactment of nonuniform digital signature laws in each of the several states will negatively impact the growth of electronic banking.

PRIVACY

There are two central issues with respect to privacy in Internet financial transactions -- preserving the confidentiality and integrity of online financial transactions and limiting disclosure of personal financial records to third parties. The transmissions involved in online financial transactions are protected, to some extent, by encryption technology.Third-party access to financial records remains a major concern.

Federal law limits access to certain financial information. For example, the Financial Records Privacy Act (12 U.S.C. § 3402) limits the conditions under which government authorities may access customer records held by financial institutions. Federal law currently prohibits creditors from gathering certain kinds of information (e.g., race, color, religion, national origin, childbearing plans) about debtors. Equal Credit Opportunity Act, 15 U.S.C. § 1691 et seq. Yet online merchants and service providers are increasingly attempting to secure and store just such data in order to better advertise and sell to their target markets.

Questions naturally arise over privacy issues such as who bears the responsibility for accuracy of personal information, and should have access to the trail of information left by online purchases. As banks move to cut out middlemen such as CyberCash®, DigiCash, and First Virtual, they may run afoul of existing statutes and regulations which limit their ability to accept and transfer certain data which comes their way in the ordinary course of business conducted over the Internet. Similarly, if banks begin to freely collect and exchange sometimes unreliable Internet transactional data, they may expose themselves to increased liability under various state and federal consumer protection statutes, such as the Fair Credit Reporting Act, 15 U.S.C. § 1681 et seq., which governs the collection and use of personal data by credit reporting agencies.

CONCLUSION

It is still not clear whether the Internet will be a bank-friendly environment. The competitive advantage of banks in facilitating commercial and consumer financial transactions is fast being eroded. Not only will it be difficult for banking security to resist the onslaught of cybertheft and cyberfraud in an Internet environment -- but there is a substantial risk that traditional banks which fail to adapt to the changes being wrought by the Internet will become cyberfailures. Still, the essence of banking -- what distinguishes it from the role played by an automated clearing house -- is personal service and professional judgment for customers large and small. If bankers can find a way to use the Internet to maintain their competitive advantage in these areas, they are more likely to thrive in the 21st century.

Home Banking without Breaking the Bank

Title: *Home Banking without Breaking the Bank*
Author: *CFI ProServices, Inc.*

Abstract: *One of the most powerful attractions of home banking lies in its inherent support of database marketing. Savvy financial institutions gather customer information via the home banking site. By analyzing this information, they can take advantage of a number of cross-selling opportunities.*

Biography: This paper was prepared by CFI ProServices, one of the nation's leading providers of home banking software. CFI was one of the first vendors to offer a home banking system to banks and credit unions, beginning in 1993, and is now offering its third-generation, Web-based product. More than 230 customers and one million users already rely on CFI's home banking solution, Encore! ® Personal Branch®.
Home banking is just part of CFI's commitment to helping financial institutions build equity in their customers through the benefits of total system integration. With industry-leading solutions ranging from lending systems to tightly integrated sales and service delivery systems, we bring unique expertise in how home banking can affect an institution's overall business processes and results.

Home Banking without Breaking the Bank

With nearly 40 million users and an annual growth rate of more than 50%, the World Wide Web is increasingly recognized as a powerful delivery channel for financial institutions. When integrated within an overall delivery system, online home banking can be a relatively low-cost channel that empowers customers to access their accounts and make transactions any time of the day from home or other remote locations.

A survey conducted in part by US Banker magazine finds that "customers who bank with personal computers rate online banking as the delivery channel that is most important to them, accounting for 37% of their transactions." The study indicates that once consumers try PC banking, they are hooked on its convenience.

However, for many smaller and mid-size financial institutions, the prospect of implementing a home banking system is daunting. There are fears about limited technical and financial resources. There are concerns about investing in technologies that may become obsolete. There are worries about the inability of home banking to accommodate such core businesses as loans and deposits.

Web technology and terminology confuses institution management, staff and customers alike. How can a financial institution implement and market such a revolution as home banking using existing staff and processes, and handle customer questions about technical or service issues, without jeopardizing its traditional businesses and delivery channels?

These concerns are legitimate, and the issues they raise are certainly not trivial. But these days it is clear that the benefits of developing a home banking channel are not only significant, they are quite possibly essential to an institution's competitiveness in the market.

Tapping into explosive growth

According to the 1997 Remote Electronic Banking report by the Mentis Corporation, "Currently, 28% of all banks with more than $250 million in deposits offer some form of home banking to their customers. By the end of 1998, almost one-half (47%) of these banks plan to offer a home banking program."

By the year 2002, home banking will accommodate more than 15 million customers and 22 million accounts, reports The Tower Group. During the next four years, U.S. banks will invest millions of dollars in channel management, online services, security and software integration to develop and refine their home banking presence.

Already, the attractions of home banking, for financial institutions and for their customers, are proving substantial especially since the first consumers to take to home banking are among the most desirable customers. Nearly one-quarter of Web users possess annual household incomes of more than $80,000, and half of all Web users held professional or managerial positions.

Targeting these middle-income Americans for home banking is both timely and on target, because home banking can help provide them with financial services that are now only available to the wealthy. As the cost of personal computers continues to decline, more and more middle-income consumers will get connected on line, further expanding the potential of home banking.

The key to frequent customer contact relationship management

Home banking is becoming the lynchpin of many institutions' new customer-oriented strategies. Institutions of every size are finding that home banking capabilities enable them to improve relationships with existing customers, attract and retain more customers, enhance cross selling, and provide more impressive, personalized services. Moreover, smaller institutions are discovering that unlike other "cutting edge" technologies such as screen phones and proprietary PC-based products, home banking systems fall well within the range of affordability while putting them on a competitive footing with larger institutions.

- Online banking promotes new account relationships, increases the number of accounts per household and translates into higher account balances per household. For example, of the first 5,000 customers who signed up for PC banking at Star Bank in Cincinnati, 3,600 of them were new customers. Moreover, these customers tended to consolidate their finances with Star Bank, maintained larger account balances and invested in more products.
- Even smaller financial institutions with relatively limited customer bases are discovering the increased selling opportunities and growth potential from online banking. In just five months, Marion Polk School Credit Union in Salem, Oregon ? a $145 million asset institution ? saw its online customer base increase to more than 1,000 users who generate more than 5,000 transactions per month.
- New business aside, every transaction diverted from the branch to the Web

can represent a $1 savings. The savings is even greater every time a consumer initiates an electronic payment instead of writing a check.
- Home banking gives financial institutions an overview into their customers' spending habits. Institutions can leverage this information to cross-sell customers on various products.
- As the alternative to physical branch extension, some home banking systems may reduce costs by a factor of 10. The Web inexpensively stretches a financial institution's geographic reach and potential market. Eventually, your home banking system can evolve to provide all of the services traditionally available only at the branch new account openings, loan applications, and mutual fund transactions, to name a few.
- Finally, any bank can strongly differentiate itself against competitors with a strategically crafted home banking program. Building a uniquely friendly and serviceable Web site can be as much a matter of better planning and creativity than of greater monetary investment.

Capitalize on what your customers want
Here is what research shows consumers want in terms of home banking options. They are, in order of preference, 24-hour access, affordability, ease of use, high reliability and security.

This indicates that today's consumer is less focused on banking products than on the components of better service. Enhancing relationship management is an essential reason for offering home banking. Retaining customers for the long term especially emerging affluent customers is a primary goal. Many institutions, particularly credit unions, are not even charging fees for their home banking services, considering it a worthy investment in relationship-building.

Although home banking is most common among big banks, by year-end 1998, it is estimated that more than one-third of even the smallest institutions plan to offer some sort of home banking. However, what few realize is how to implement or manage this type of program.

1998: The Year of Systems Integration
Until recently, financial institutions implemented home banking systems primarily to advance their strategic intents, including generating fee income, building customer relationships, selling products, expanding delivery channels and driving down costs.

Today strategic intents remain important, but the focus now is integrating delivery channels to achieve the business objectives. In some cases, home banking is still set up as a separate channel and may be the only system for bringing together all of a customer's accounts. Usually this represents an interim solution, while connectivity to other retail delivery systems and comprehensive relationship management tools are built.

One of the biggest challenges facing institutions in 1998 is to move beyond just supporting simple account inquiry capabilities at their Web sites. They are going to have to offer integrated whole-banking functions, such as loan and new account transactions, and bring full relationship banking to their Web sites to cross-sell and up-sell to customers.

A key factor in selecting a software contractor is determining how well it can integrate home banking with the other delivery channels. The less an institution needs to change its existing systems, the better. Certainly, home banking becomes extremely attractive when it can help move banks from an isolated to an integrated delivery strategy.

Financial institutions will need to make adjustments to implement home banking, but the good news is that in many cases, key processes that are already in place won't need to change.

How to change less but make the most of change

To ease into home banking, most small institutions can and should look to third-party providers to help them meet the major implementation challenges. For example, rather than begin by hosting their own sites, many small and mid-size financial institutions will prefer to partner with a service bureau.

This reduces start-up costs, takes a step-by-step approach to product and service delivery, and enables an in-house system to gradually take hold. Most service bureaus offer access to Web site developers, provide technical Web service to customers, house the Web site and online banking software, and provide host communications links.

Still, some adjustments will have to be made in the way the financial institution does business. For example, banks may wish to add systems that complement the non-personal Web environment. Banking staffs may have to adjust to working on screen rather than on paper. And institutions will need to change how documents are delivered to a customer, once a relationship is established.

Actual channel management costs can vary greatly among institutions. It is likely, however, that channel costs per customer will decline significantly over the next five years, as the rapidly growing number of online customers creates economies of scale. Since most aspects of home banking can be outsourced, and institutions can contract for variable fees based on active customers, this rapid cost reduction is also partly based on the maturity of available support applications, combinations of delivery channels, and the ability of service bureaus to pass their economies of scale on to customers.

One possible change well worth considering at this watershed era in financial services: institutions currently utilizing a proprietary system should seriously consider switching to an open one. According to the 1996 Bank Director's White Paper on Retail Technology, there are five major advantages in moving to an open systems environment:

- Institutions can take advantage of lower costs.
- They can enjoy easier system expandability and scalability.
- They can choose from a wider and more competitive range of options.

- They can more readily establish standard platforms across the corporation.
- They have greater independence in selecting vendors.

The impact of these benefits will be all the more significant in an online world.

Back-office systems where data for transactions is already defined need not and should not be redefined to fit any particular Internet banking model. The goal should be to provide customers with real-time online access to data. Many early implementations of home banking included daily batch file transfers, which accommodated most servicing needs while providing useful buffers for security management. However, with the development of more advanced security systems, the goal for institutions today is to provide real-time processing and the full integration of call center and home banking programs. Utilizing an open system architecture greatly simplifies this challenge.

Alternative data collection methods, such as Internet-driven new account opening, should connect to existing back-office systems using industry standard communications and messaging protocols available through middleware. While customer data may need to be cleaned and standardized, the institution should then be able to leverage this same data for outbound telemarketing, financial planning, and cross-selling via home banking. For example, with clean customer data, institutions can target customers by geography or other demographics and directly market products and services to specific individuals.

Transferring transactional processes to the Web

A home banking system can provide the opportunity to revamp processes for taking loan applications and opening new accounts. Institutions need to be thinking about a whole-banking strategy: how to open deposit accounts, apply for loans, perform credit checks and actually close loans online?not just do account inquiries.

However, not all phases of loan and new account transactions can currently be handled online. Even though customers can currently receive tentative confirmation, original contracts and agreements must still be exchanged by mail to ensure that no modifications are made to the printed form. Leading Internet banking system vendors are automating regulatory disclosures for online delivery and streamlining processes to minimize the need for manual intervention.

In the meantime, financial institutions must consider how closing documents are completed and exchanged. At Commerce Bank, Cherry Hill, N.J., online loan prospects are directed to the bank's loan-by-phone, where a personal banker quickly follows up and closes the transaction. With this approach, Commerce Bank processes as many loan applications online as in a typical branch. The bank has found that this system reduces paperwork, speeds credit decision making and lowers transaction costs, while maximizing convenience to the customer.

Coming to grips with compliance

Financial institutions need to partner closely with legal experts, who can guide them past regulatory issues involving home banking. Laws and regulations applicable to the Web already exist and are evolving. According to Alan Dombrow, senior manager and regulatory liaison at CFI ProServices, and former National Bank Examiner of the OCC, "While there are no separate regulations directly addressing compliance issues for Web sites, regulators have said they view Web pages as advertisements with all the incumbent regulations.

"The Federal Reserve Board believes that consumers receiving disclosures by electronic communication should have protections regarding delivery similar to those afforded consumers receiving disclosures in paper form.

"Home banking presents many regulatory challenges to regulators and to those in the financial services industry. The goal of clearly informing consumers about their rights and obligations remains constant as delivery channels evolve."

To fee or not to fee?
When and how much to assess customers for home banking is a question dependent on each institution's pricing strategies. More and more financial institutions are offering home banking for free and charging modest fees for electronic bill payment. Some institutions price their products according to the type of delivery channel. Others have found that customers are willing to pay a premium for certain online products as the cost of greater convenience.

Marketing assistance part of the package?
One of the most powerful attractions of home banking lies in its inherent support of database marketing. Savvy financial institutions gather customer information via the home banking site. By analyzing this information, they can take advantage of a number of cross-selling opportunities.

When it comes to launching and publicizing the Web site itself, however, small to mid-sized financial institutions can face an uphill battle, since they often lack the marketing budget necessary to adequately promote these services. It is the mark of a complete and professional home banking systems provider that its package includes marketing support tools?including lobby posters, statement stuffers, and discount Internet access?to help lure the institution's customers online.

A checklist for launching
The home banking channel should be seen and understood as an integral part of a financial institution's total suite of customer services. Institution personnel should be versed in the services it offers and understand how to help customers learn more.

Before launching a home banking program, experts suggest holding training sessions for end users as well as customer service personnel. Some institutions identify branch-based PC specialists who can provide one-on-one local service, or contract with third-party sources to provide toll-free telephone technical support to users.

Here are some basic rules to remember as the system prepares to go public:

- Establish and think through all the new processes that direct banking entails.
- Make sure that the technology is sound and can handle growth.
- Involve and train the CSRs--they're vital to providing a positive customer experience.
- Roll out the service to your most Web-savvy customers first.
- Prepare for customer feedback via multiple channels and establish a system to deal with it.
- Remember that the system won't be perfect the first time, so keep money in the budget for revisions.
- Keep users immediately apprised via e-mail of changes and new developments.

Research shows that once users gain familiarity with and confidence in online banking, it becomes their delivery channel of choice and that opens new avenues of cross-selling, cost-savings and convenience that work to the benefit of all.

Moving toward the 21st Century

On the cusp of the new millennium, online banking is ushering in an entirely new culture--for financial institutions as well as for their millions of customers. The ability to transfer funds, pay bills, apply for loans, and open new accounts is just the tip of the proverbial iceberg.

In the not-too-distant future, institutions will add new delivery channels, including banking kiosks, television-based Web access, screen phones and interactive television. As they integrate multiple delivery channels, they will benefit greatly from an open systems environment. It is increasingly crucial that institutions choose partners who can facilitate these long-term technical goals.

The Bank Director's White Paper on Retail Technology calls the financial services community to action to put this two-edged sword in their arsenal of tools: "Banks must recognize both the threat and the opportunity of the electronic future. They must act quickly to extend their unique experience and insight to the medium of electronic commerce."

Bringing Remote Banking into the Mainstream

Title: *Bringing Remote Banking into the Mainstream*
Author: *ACI*
Abstract: *To the financial institution, remote banking has evolved to encompass many forms of direct customer access to a broad range of financial products and services. These can be as simple as requesting an account balance or paying a bill or as complex as applying for credit or initiating a stock trade. To meet the needs of particular market segments, financial institutions have access to a variety of technologies that could be used to offer remote banking services to their customers.*

Introduction

Remote banking, direct banking, home banking—whatever term you prefer, each means different things to different people. A visit to an automated teller machine (ATM) may be considered remote banking by some customers, while others see remote banking as the ability to access financial services using a broad variety of technologies. U.S. customers may choose to pay their bills over the Internet, while customers in Japan may elect to visit an ATM to pay their bills. Customers of Brazilian banks expect their ATMs to dispense personal checks, while customers in the U.K. are accustomed to cashing checks at self-service kiosks. Today, customers all over the world expect to access their banking services using a wide variety of delivery technologies delivery channels.

To the financial institution, remote banking has evolved to encompass many forms of direct customer access to a broad range of financial products and services. These can be as simple as requesting an account balance or paying a bill or as complex as applying for credit or initiating a stock trade. To meet the needs of particular market segments, financial institutions have access to a variety of technologies that could be used to offer remote banking services to their customers. Access devices include everything from ATMs and self-service kiosks to telephones and personal computers. Delivery channels may include ATM networks, Web servers, and call center systems, as well as interfaces to service providers like Visa Interactive, Integrion, CheckFree and others.

The challenge financial institutions face today is to structure their remote banking services to provide high levels of service and information integrity while accommodating a multitude of access devices and delivery channels. This can be difficult, given the vast selection of service solutions available in today's market. Part of the challenge is that the information and services customers want to access reside on a wide variety of applications and

Figure 1

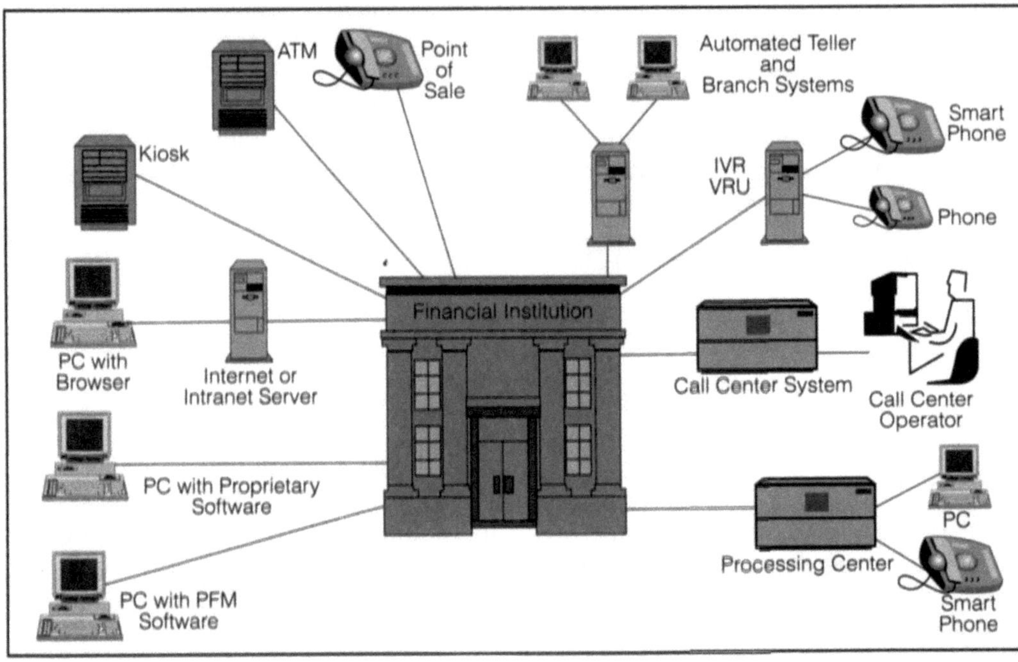

systems, some maintained internally and others outsourced to third parties. Compounding the challenge, financial institutions must consult their crystal balls to look beyond the needs of today's customer and anticipate the future. Remote banking solutions need to be designed with a high degree of flexibility today, in order to meet the needs of customers tomorrow.

Remote Banking: Creating The Ideal Solution

StrategiesService Package

Depending on local market competition and the available resources, financial institutions have tried a variety of approaches to remote banking.

Some have adopted a wait-and-see attitude, allowing other financial institutions to make the initial investment and expend the resources required to test the consumer market. In this case, the financial institution may plan to enter the market once successful, proven remote banking solutions emerge.

Other financial institutions have decided on a phased-in approach. The first step is usually to bring a product to market as quickly as possible in order to meet the immediate demand and to protect the customer base against aggressive competitors. Here, the financial institution may initially offer a basic remote banking service package providing bill payment services and access to checking and savings information via telephone and/or PC. At a later date, they may move to upgrade and expand remote banking functionality and access methods as market demand dictates.

Some financial institutions have taken the shotgun approach to remote banking. With the broad range of home banking solutions available, these institutions have decided to implement a variety of applications. Here, their strategy is to make sure the customers have enough choices at their financial institution, so they have no reason to move to the competition.

Still others have implemented a comprehensive, high-capability remote banking strategy. In this approach, the institution provides both retail and commercial customers remote access to information and services on checking, savings, lending, credit, and investment products. The customer is provided entry via a wide range of access devices and a variety of channels including the Internet.

Comprehensive Solutions

No matter what initial approach is taken, most institutions would agree that long-term offerings must be comprehensive. The ultimate vision of remote banking should include most, if not all, of the following attributes:

- Support for a wide variety of access devices and delivery channels, including the Internet
- Support for proprietary and/or commercial Personal Financial Management (PFM) software such as Intuit's Quicken, MECA's Managing Your Money or Microsoft Money
- 24-hour/7-day-a-week service availability
- Real-time access to current and accurate customer, account, and financial information
- Ability to scale the offerings as subscribers and transaction volumes increase
- Incorporation of existing delivery channels such as ATM, point-of-sale (POS), and teller, so that the customer view of account information is consistent across all channels
- Ability to bring products to market as quickly as possible to meet current consumer demand
- Flexibility to add new services and delivery technologies as customers demand
- Modular design to cost-effectively meet long-term needs
- High degree of transaction and data integrity and security
- Flexibility to add new services and delivery technologies as customers demand
- Infrastructure that leverages existing investment in legacy systems
- Flexibility to integrate service provider technologies with the technologies currently owned by the financial institution in order to offer a complete remote banking service package
- A solution that allows the financial institution to differentiate itself and build customer loyalty

Deployment Considerations

Meeting the objectives outlined above is becoming an increasingly complicated and costly task that can hamper a quick market entry. In addition to acquiring new delivery technologies, integrating them into existing financial systems can require:

- Modifying existing systems to provide network connectivity needed to support PCs, IVRs, the Internet, PFM software and other technologies that may be deployed
- Developing new application software to process requests and transactions from new access devices and delivery channels
- Modifying existing applications and systems to integrate solutions so that financial institution customers receive consistent information across existing and new access devices and delivery channels
- Building interfaces to external service providers to support the various customer products and services a financial institution chooses to outsource
- Building or purchasing new application software to provide complicated remote banking and bill payment services like warehousing future-dated bill payment transactions, maintaining and managing "payee" lists, authorizing funds availability, etc.
- Building and maintaining specific interface software to support messaging formats required by particular service providers and delivery technologies

> "A bank's ability to integrate its content with its navigation will be dependent on the current legacy systems. Upgrading these will require an investment that is several orders of magnitude greater than that required to create the front end."
>
> *Boston Consulting Group. The Information Superhighway and Retail Banking*

Industry experts all agree that innovative remote banking efforts will require integration of new technologies into existing infrastructures and systems. These efforts will cost institutions a great deal in time, money, and resources. In fact, the resources required for integration may exceed the initial cost of the new technology itself.

While the initial technology investment alone can be large, long-term operations and maintenance costs of interface and application software residing on multiple systems can also skyrocket. Without an integrated strategy that crosses all delivery channels, many institutions will suffer the brutal costs of maintaining an environment where multiple delivery devices must connect to multiple back-end systems using multiple interfaces.

If all delivery channels do not utilize identical customer account information, customers will see a different view of their current account relationships depending upon the device or delivery channel used. This potentially serious product weakness would undermine customer confidence and acceptance of the service.

As technologies change and customers demand remote access to more products

Figure 2

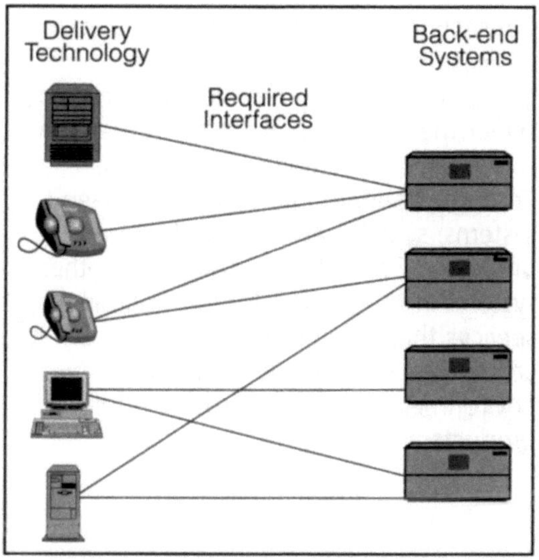

and services, it becomes increasingly difficult to respond in a cost-effective manner. With each new delivery technology added, a new interface to the processing system must be established, making it even more difficult to ensure that consistent account information can be accessed across all delivery channels. Before long, this type of environment becomes unmanageable and cost prohibitive to maintain.

The BASE24 Alternative

ACI's BASE24 remote banking solutions are designed to address the most complex and costly aspect of implementing an end-to-end remote banking solution—integrating new delivery devices and channels in a way that minimizes modifications to existing systems and does not require developing new back-end systems. BASE24 enables financial institutions to quickly and cost-effectively integrate and support new delivery technologies, such as PC software, web servers, IVRs and other technologies. ACI achieves this in a way that provides clear advantages, accepted and proven by some of the most demanding electronic financial service providers around the world.

Figure 3

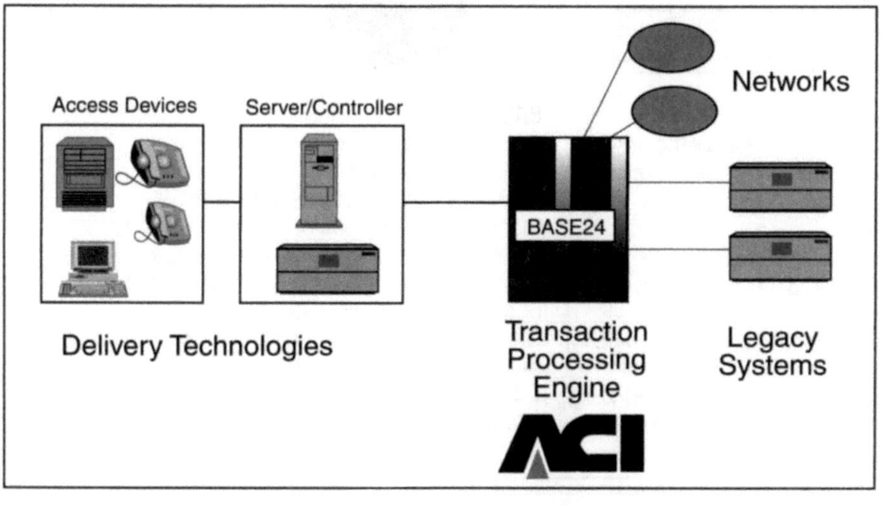

As depicted in the diagram, end-to-end remote banking deployment consists of several components of disparate hardware and operating systems. Each component in the configuration plays a complementary role best suited to the strengths of that particular component.

Delivery technology consists of customer access devices such as ATMs, telephones, PCs, and screenphones, while the server/controller such as the IVR or web server provides the presentation to the customers (i.e., screens and voice scripts). ACI's BASE24 system acts as the engine that can authenticate customers, provide real-time customer account, statement, and vendor information, and route and authorize financial and bill payment transactions generated by remote banking devices. In effect, it acts as a high-performance, highly reliable, fault-tolerant server that processes financial transactions for delivery technologies of all types. Through standard interfaces to back-end systems, switch networks or third-party processors, BASE24 can also access other systems and networks for information and services that may not reside on BASE24. It creates the cooperative computing environment needed to provide a full set of products and services.

Figure 4

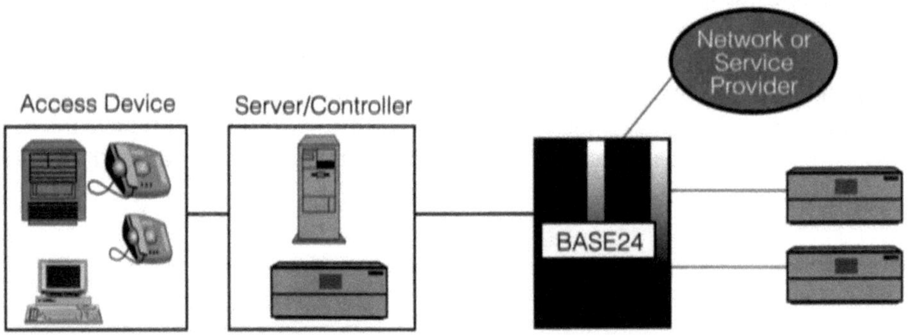

Access Device
Used by the customer to access the service
Receives "presentation" from the server

Server Controller
Contains the customer view or service "presentation" (screens, voice scripts, navigation options)
Requests financial information or transaction processing from host financial systems
Formats responses (screens or scripts) with appropriate information back to the customer device.

BASE24
Authenticates the customer and financial transaction request
Maintains balance and history database to authorize transactions or gather information from appropriate hosts or external systems
Can route transactions to appropriate networks or systems for external processing
Creates transaction log and maintains database and transaction integrity
Communicates appropriate financial information to server

It creates the cooperative computing environment needed to provide a full set of products and services.

When the delivery channels need customer-specific information such as an account balance or statement information, or needs to authorize a financial transaction such as a funds transfer or bill payment, they communicate the request to the BASE24 system. BASE24 fulfills the request and/or routes and authorizes the financial transaction and provides information back to the delivery channels, which then communicate an appropriate response to the customer via the access device.

If the delivery channels need to access data or services that do not reside on the BASE24 platform, BASE24 can intelligently route requests to other systems, gather information from those systems and then forward responses to the delivery channelomponents for presentation to the customer. In this way, host systems are modified to communicate only with the BASE24 system, and do not have to be programmed to address each individual server, controller, or access device. BASE24 minimizes and simplifies the interface issues and programming on the host systems.

Using BASE24 as the engine that can authorize, route, and process financial transactions and requests generated by any remote delivery technology offers the following advantages:

- Reduces time to market and deployment costs by greatly reducing in-house software development—there is no longer a need to write multiple interfaces between access devices or delivery channels and each back-end system
- Allows the integration of all forms of electronic banking (ATM, POS, teller, phone banking, PC banking, PFM software, Internet banking, and others), so that remote banking customers can receive consistent information and services across all technologies
- A standard message interface that enables other vendor technologies (i.e., Voice Response Units, Web servers, etc.) to easily communicate with BASE24 - remote banking applications
- Provides secure, reliable, high-performance financial transaction processing, 7 days a week 24 hours a day, 365 days a year, as proven around the world in EFT networks
- Provides the scalability required to respond to growing customer demand for remote banking and the flexibility to easily support new delivery technologies, financial products, and services as they become available
- Provides financial institutions access to the knowledge and expertise ACI has earned through worldwide deployments of remote banking solutions

The Value of ACI's BASE24 Solutions

Connectivity:
Enabling delivery technology to communicate to existing systems within the financial institution is an absolute requirement for the real-time processing of remote banking transactions and inquiries. The types of devices and networks customers will use to access financial

Figure 5

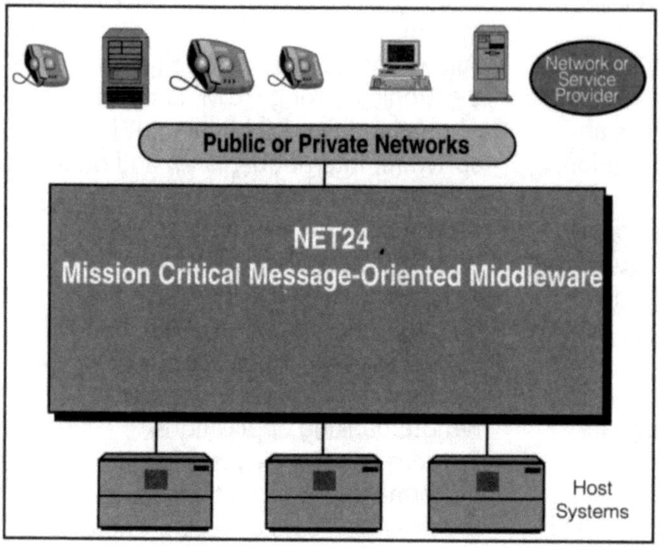

others, to provide customers with the products and services they desire. Whether the financial institution needs to create a cooperative computing environment with existing SNA, bisync or X.25 networks, or a network like the Internet that uses TCP/IP protocols, BASE24 solutions will allow the institution to save time and development costs associated with building mission-critical, scalable, and flexible networking capability.

information and services will vary widely. The wider the variety of entry points, the greater the pressure on existing systems to support the telecommunications protocols necessary to connect with a wide range of networks. Modifying applications on existing systems to support the technical requirements of public networks (like the Internet), proprietary networks, (such as those accessed by ATMs), card-switch networks, in-house local area networks, and existing system networks can be costly and time consuming.

BASE24 solutions are built on message-oriented middleware called NET24. NET24 supports all the protocols necessary to communicate with any delivery technology the financial institution chooses. NET24 also supports communications with existing systems including demand deposit (current) systems, credit systems, loan systems, and

Comprehensive Operations and Monitoring: NET24 provides a single point of monitoring and control across the entire network including connections to all forms of delivery mechanisms, back-end systems, third-party service providers, switch networks, and distributed BASE24 systems. This single view of the overall service operation enables the financial institution to ensure that all delivery channels, network connections, and processing systems are operable.
Having a single point from which computer and network monitoring can be performed is crucial to providing the highest level of service required in today's competitive environment. Without this capability, it is possible that a PC banking customer may attempt to log on to a financial institution's remote banking service, and even though the server is operable, the customer cannot connect because the institution is unaware

that its modem bank is down.
NET24 provides a single point of monitoring and control across the entire network including connections to all forms of delivery mechanisms, back-end systems, third-party service providers, switch networks, and distributed BASE24 systems.

EFT Fundamentals:
ACI strongly believes mission-critical fundamentals, such as those demanded by EFT systems around the world, are prerequisites in the worldwide remote banking environment. Remote banking customers will expect reliable service where they want it and when they want it. BASE24 applications, which run on the Tandem NSK platform, deliver the following benefits demanded in the financial services environment:

- 24-hour availability
- Linear scalability of the platform to grow as the customer base and usage of services grow
- Highest levels of transaction and data integrity, ensuring that transactions are not lost or processed incompletely
- Fault-tolerance to isolate the impacts of system outages from the customer
- High-performance transaction processing to ensure very quick response times
- Ability to handle extremely large transaction volumes and high peak loads that will occur given the nature of the services being offered

ACI's BASE24 solutions are relied upon to deliver these fundamentals to traditional EFT networks worldwide. BASE24 remote banking solutions incorporate what ACI has learned from the EFT world, applying the same high standards for mission-critical financial applications to the world of remote banking.

Financial Services Applications:
ACI's family of BASE24 applications are built on an integrated architecture that allows institutions to run multiple payment and delivery applications from a single platform. Sharing common databases and common networking facilities allows institutions to delivery a variety of consistent products, services and information across any access device they choose to deploy and support. When

Figure 6

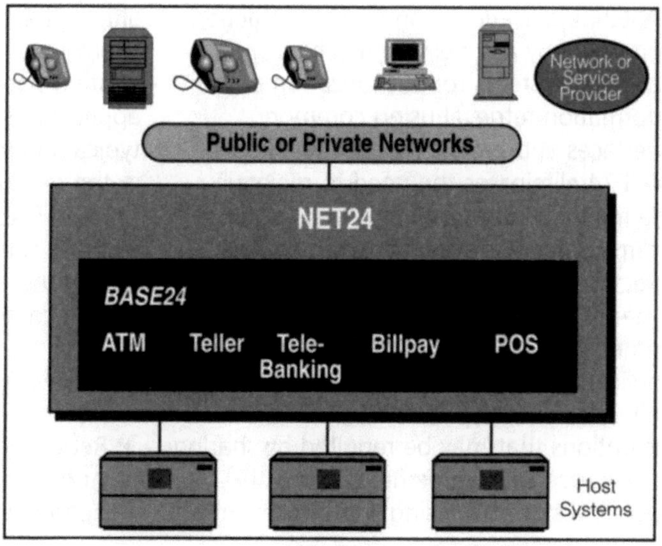

accessing an institution's remote banking service, it is essential that customers receive consistent information and behavior across all delivery technologies. Remote banking solutions that depend upon multiple financial services applications must offer an appropriate level of database integration.

BASE24 is a family of solutions that run under a common networking platform (NET24). NET24 provides connectivity to and accepts transactions and requests from multiple sources and delivers those transactions to appropriate applications for processing.

Through this common application platform and integrated family of applications, financial institutions can ensure their customers receive access to real-time information and transaction processing that is consistent across any means chosen to access the financial institution.

BASE24 and NET24 create a cooperative computing environment that leverages existing systems to deliver financial services through new remote banking channels. BASE24 applications can route transactions generated by any delivery mechanism to back-end systems for authorization and information retrieval using common interfaces and messaging standards. BASE24 eliminates the need to maintain the multiple interfaces that often result from connecting a set of access devices directly to another set of host back-end systems.

In the same fashion, ACI's platform and applications can interface to switch networks, service providers, and other institutions that may be required by sharing or co-processing agreements. As with back-end systems, a single interface can be used no matter how the transactions are generated—greatly simplifying the networking environment and reducing maintenance costs.

Remote Banking Service Applications: Responding to consumer demand for increased banking and financial management capabilities, financial institutions must find a way to support this expanded service by offering constant availability and access to consistent information using a variety of devices and channels. Remote banking services are expanding to include real-time, 24-hour-a-day access to customer and account information, transaction authorization, and processing. Most existing financial service software applications were not typically not designed to support the customer-initiated transactions and requests that will originate from the new generation of remote banking services and delivery technologies.

ACI offers BASE24-telebanking and BASE24-billpay applications as an alternative to internally developing these capabilities and incorporating them into existing back-end systems. These applications support real-time servicing of typical financial transactions encountered in the remote banking environment.

ACI's BASE24-telebanking application includes the functions traditionally performed at the branch or by a customer service center, including:

- Reviewing past transactions and account history
- Requesting balance inquiries
- Authorizing immediate transfers between intrabank accounts (immediate,

future and recurring)
- Changing or verifying a PIN
- Customizable service options such as submitting credit or loan application requests and ordering personal checks

ACI's BASE24-billpay application allows institutions to offer full bill payment capabilities to their customers. The application allows financial institutions to maintain and manage open and closed vendor lists, warehouse future-dated payments and transfers, authorize funds availability prior to payment approval and generate remittance data. All of these functions are required in order to offer customers the following capabilities:
- Authorize immediate, future, and recurring bill payments
- Authorize immediate, future, and recurring interrabank transfers
- Authorize future and recurring intrabank transfers
- Perform vendor management functions including vendor list inquiries
- Process scheduled payments and transfers inquiries
- Delete scheduled payments and transfers
- Change scheduled payments and transfers
- Process last "X" payments and transfers inquiries

ACI's remote banking solutions provide these capabilities consistently across the entire set of delivery options currently supported by institutions today. Transactions and requests generated by customers using telephones, personal computers, call centers and personal financial management software, or by service providers such as Visa Interactive, Integrion, and others, can all be part of the remote banking solution using ACI's BASE24 products.

Open Support for Delivery Technologies: ACI strongly believes financial institutions must be able to choose delivery technologies from vendors best suited to meet their needs.. That is why BASE24 Remote Banking was developed using open industry standards that enable any access device or delivery channel to easily interface with BASE24 applications. In addition, ACI maintains a number of strategic partnerships with key vendors in the industry who already support an interface to BASE24 applications. This open support philosophy enables institutions to maintain control of their delivery strategy rather than be forced into one technology over another.
Through messaging standards, BASE24 can accept and process transactions from virtually any technology provided by any vendor or developed by the financial institution, including support for:
- Delivery products based on common industry platforms such as UNIX, Windows NT, OS/2, DOS, Tandem NSK, and others
- PC products like Microsoft's Money, Intuit's Quicken, or software authored by the financial institution or a third party
- Service provider systems operated by Integrion, VISA Interactive, CheckFree, and others
- Industry standard Internet browsers and Web server technologies from a variety of vendors

Incorporation of New Technologies:
ACI is a leader in providing innovative financial transaction processing solutions. Some of our most recent activities have involved incorporating smart (chip) card-based technologies, such as stored-value and chip-based authentication of debit and credit cards, into our core applications.
As the industry adopts smart card technology to further secure remote banking transactions (including those on the Internet), ACI will continue to incorporate these capabilities into its remote banking solutions. And as other technologies are developed, financial institutions can depend on ACI to enable them to take advantage of these technologies.

ACI Means Proven Financial Systems:
ACI is a recognized leader in the provision of integrated transaction processing and payments solutions. ACI applications are feature-rich, address the wide range of traditional and emerging forms of service delivery, and support user configuration to meet specific requirements.
Our products are built on networking foundations that feature manageability through an integrated systems environment, extensibility to support multiple applications, and flexibility to operate efficiently in complex network environments. Consistent with ACI's rigorous product mindset, our solutions are designed to interoperate, provide continuous service availability, ensure information integrity, and scale to meet growing transaction volumes.
Financial institutions with existing ACI front-end systems such as ATM and POS, can leverage their knowledge of and experience with ACI applications in order to ease the integration of existing systems into a comprehensive remote banking environment.

Service and Support:
Superior ACI customer service and support comes with every product including comprehensive project management, hardware preparation, installation and go-live support, documentation, training, 24-hour hotline and ongoing account management.

Summary

Whether a financial institution has yet to begin its remote banking initiative or is well into the implementation of a complex remote delivery service strategy, ACI's BASE24 telebanking, billpay, and NET24 products are proven, reliable solutions. Installing the BASE24-telebanking and BASE24-billpay applications will:

- Enable the institution to choose the delivery strategy best suited to its needs and the needs of its customers—allowings the financial institution to maintain control of its electronic financial services offerings
- Minimize the amount of software modifications to existing systems,—saving time, money, and risk introduced by making changes to legacy systems
- Provide common interfaces throughout a financial institution's networking and computing environment—Minimizing ongoing maintenance costs and the costs of expanding services by providing common interfaces throughout a financial institution's networking and

computing environment
- Create the cooperative computing environment necessary to offer cost-effective, revenue-generating, full-service features to customers
- Provide the NET24 application middleware vital to supporting network connectivity between a variety of access devices and delivery mechanisms, and existing legacy systems and service applications—creating the cooperative computing environment necessary to offer cost-effective, revenue-generating, full-service features to customers
- Enable customers to access customer and account information that is consistent across all existing and new delivery channels and supported in a real-time environment—instillings customer confidence in the service
- Provide 24-hour availability, support high transaction volumes, and ensure the integrity and reliability of remote banking services
- Provides the scalability and flexibility required to accommodate increased customer usage—maximizing use of current technology while minimizing ongoing expenses to support increased volume
- Provide monitoring and control of all delivery mechanisms from a single network management facility—resultings in the highest level of overall system reliability and efficiency

For over two decades, ACI has been a leading provider of financial transaction processing solutions utilized by organizations all over the world. Helping financial institutions deliver integrated, highly reliable, high-performance, cost-effective, flexible, and innovative financial services is our goal.

ACI's solutions are designed to help financial institutions meet the remote banking objectives and characteristics explained in this document. From our interaction with financial institutions around the world, we see the criteria listed above as long-term critical success factors for the delivery of electronic financial services.

ACI is dedicated to helping more financial institutions successfully adopt remote banking as a mainstream financial service delivery technology.

Transforming Consumer Banking Through Internet Technology

Title: *Transforming Consumer Banking Through Internet Technology*
Author: *USWeb Corporation*

Abstract: *A USWeb study of 1,000 of the largest financial services companies in the United States (all with annual sales of at least $250 million) revealed that 93% of banks plan some type of Web development within the next few years. The question for the banks, thrifts and other financial services institutions that have yet to incorporate the Internet into their business plans has shifted from "Should we get online " to "How should we get online?"*

Biography: USWeb Corporation, headquartered in Santa Clara, California, is a leading international professional services firm dedicated to helping businesses solve their mission-critical objectives through Web-based technology. (more at the end of the article)

Introduction

Internet Banking Comes of Age
For more than 15 years, industry experts have predicted that home electronic banking would finally reach a critical mass of consumer acceptance and that it would soon be commonplace to pay bills and access financial accounts from home. And every year, the banks and financial institutions that believed the predictions spent millions of dollars developing, marketing and supporting their early home banking systems. The investments they made weren't just the multi-million dollar expenditures for the custom consumer software applications that many developed. They also had to create massive new IT departments to design and support the proprietary back-end systems, and install and maintain the hundreds of dial-in lines and racks of modems their customers would use for access. Virtually without exception, these home banking systems failed to meet expectations and were soon scaled back in scope or abandoned entirely.

Today, online banking is finally a reality, both in rapidly growing consumer acceptance and in the financial results being seen by institutions that are deploying state-of-the-art electronic banking.

What is different this time around? First, the huge investments required in proprietary IT have disappeared almost entirely. With the rise of the Internet and standard TCP/IP networking, banks (1) can now offer online banking without dial-in lines, modems and all of the necessary equipment and people needed to support them. All that's needed now, from an IT standpoint, is a standard connection to the Internet, at a cost that can be as low as $1,000 per month.

Second, the development of universal standard protocols such as OFX and Integrion Gold make it easier for financial institutions to support online banking without any investment in client software development. In the past, the development of software for consumer use was one of the major expenditures in delivering an online banking solution. In addition to the costs of developing and testing software were those incurred in duplication, distribution, technical support and maintenance upgrades. Banks also had to decide if they were going to target sophisticated banking customers by including full budgeting and financial planning features. Or they could target less sophisticated users by offering only simple transaction processing, bill payment and account statement access. Either choice limited the market's acceptance of the solution.

Today, developing an online banking solution using OFX, for example, allows a bank to please both types of consumers while removing all of the costs associated with the development, distribution and support of client software. A well-designed solution will automatically allow users of the predominant consumer financial software packages such as Microsoft Money TM and Intuit's Quicken TM to access statements, pay bills and initiate transactions. At the same time, the system permits access via the Internet for customers who don't need full financial planning but want a full range of transaction and account access features such as bill payment, fund transfers and statement review. In both cases, the burden of client software support and development has shifted to either the Web browser developer or the financial planning software developer. The bank is free to focus on the critical areas in which it can add value for its customers: back-end system functionality, product design, marketing and customer support. More importantly, both the bank and its customers can begin using online banking.

(1) For convenience, this document refers to "banks" and "banking"; however, it should be understood that the issues and statements apply equally to all types of depository institutions, including banks, thrifts, savings and loans, and credit unions.

The third major change affecting the acceptance of online banking is the availability of effective and reliable transaction security systems. One of the biggest concerns about online consumer banking has always been the security of the systems and the protection of sensitive transaction data. While security remains a major issue for both the bank and its customers, there are solutions today that are cost effective and permit an unprecedented degree of security, reliability and protection from misuse.

This white paper explores the growth in online banking and the ways in which financial institutions can take advantage of Internet technology to offer cost-effective banking solutions. It outlines how the Internet is affecting banks, thrifts and financial services companies today—and what some of the opportunities will be in the future. Next, it introduces a four-phase framework for understanding the different ways banks can get online—and the advantages of each level of online commitment. Finally, it offers some suggestions for developing an Internet banking offering. The paper concludes with a few tips for implementing an effective online strategy.

Figure 1: Online Banking Usage

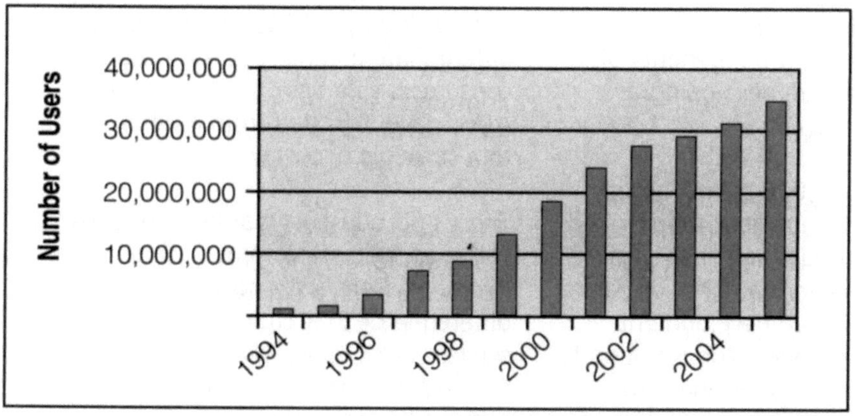

Source: Online Banking Report, January 1998

Figure 2: Number of Banks with Web Sites

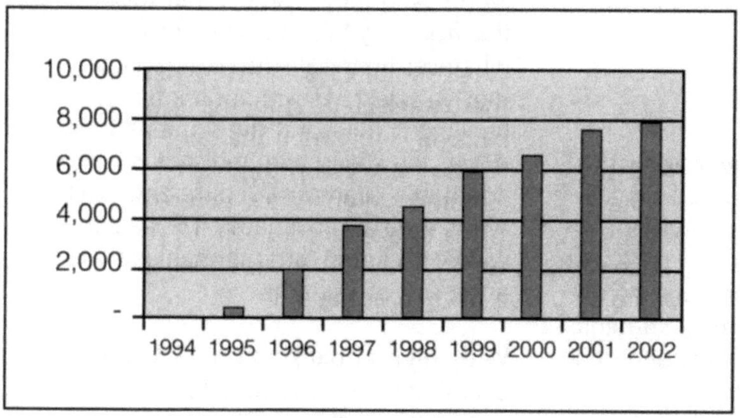

Source: Online Banking Report, January 1998

The Growth of Internet Banking

Internet banking is growing faster today than most financial services institutions had ever expected. Today, roughly 4.5 million households use Internet banking and or bill payment at least once a month, and that number is expected to increase to 33.5 million—nearly 31% of all U.S. households—by 2005 (2).

What's contributing to this growth in online banking and bill payment? First, Internet usage is on the rise. Today, more than 100 million people have Internet access worldwide, and that number is increasing daily. Second, improvements in technology and customer acceptance have propelled the role of electronic commerce. More than 10 million people have made a purchase online, and that number is expected to grow (3). And third, more banks, thrifts and other financial institutions are realizing the benefits of online banking and are developing a presence on the Internet.

A USWeb study of 1,000 of the largest financial services companies in the United States (all with annual sales of at least $250 million) revealed that 93% of banks plan some type of Web development within the next few years. The question for the banks, thrifts and other financial services institutions that have yet to incorporate the Internet into their business plans has shifted from "Should we get online " to "How should we get online?"

(2) Online Banking Report, January 1998
(3) Online Banking Report, January 1998

Internet Banking as a Strategic Necessity
Given the wealth of opportunities the Internet creates for financial services companies, and the accelerated pace with which banks are going online, having an Internet presence will become a strategic necessity for most banks, thrifts and other financial services institutions.

To understand this, consider the affect of ATMs on the banking industry. From 1977 to 1988, Citibank, an early adopter of ATM technology, increased its market share in New York City from 4% to 13.4% (4). Many analysts agree that ATMs were a substantial driver of that impressive growth. Indeed, in its early stages, the ATM was a source of strategic differentiation for Citibank and other early adopters. But, as the technology was deployed more widely, the source of value associated with having ATM technology shifted. Today, ATM technology doesn't differentiate a bank—it's expected by consumers as a basic service offering. ATMs have migrated from a differentiator to a strategic necessity.

There's no question that Internet banking will soon follow the same path as ATMs, migrating from a strong competitive differentiator to a basic and expected service. Consumers will expect to be able to check balances, pay bills, transfer funds and review transactions from anywhere, much as they now expect to be able to retrieve cash wherever they go in the world. And they'll expect that they can do this from any Web browser, or from whatever financial software package they've selected. Although Internet banking is following the same path as ATMs, the speed with which it is being adopted is dramatically different. While ATMs took approximately 15 years to be widely accepted, Internet banking will take a fraction of the time.

With Internet banking on track to become a strategic necessity for all financial services companies in the next few years, the window for making it a strategic advantage for banks is rapidly closing. For a bank to remain competitive and properly service the needs of its most attractive customer base, it must plan and deploy an online banking strategy that is consistent with its overall goals.

4 Santos and Peffers, 1995

The Opportunity for Internet Banking

The Internet poses enormous opportunities for banks, thrifts and other financial services institutions to fundamentally reshape their organizations. The benefits of the Internet permeate an organization—from marketing and sales to back office and operational functions. These benefits have caused financial organizations to view the Internet as more than a marketing communications tool and to begin to successfully employ the Internet as a new channel for their services. Some of the most relevant benefits of Internet banking follow:

Increase Customer Satisfaction
Internet banking allows customers to access banking services 24 hours a day, 7 days a week. Like ATMs, Internet banking empowers customers to choose when and where they conduct their banking.

An American Banking Association and Gallup Poll survey revealed that the primary reason customers maintain an account with a particular financial services institution is convenience. This implies that in order to retain customers today, banks need to offer their services through multiple distribution channels: physical branches, telephones, ATMs, kiosks, screen phones, PCs, and the Internet. The more delivery channels a bank offers and the more functions available on an Internet site, the more convenient it becomes for customers to conduct business—and the higher the rate of customer acquisition that a bank is likely to experience.

Expand Product Offerings
Internet banking allows financial services institutions to capture a larger percentage of their customers' asset base. Today, banks and thrifts compete with brokerage houses, insurance companies and mutual fund companies for a growing share of consumers' financial assets. The Internet allows banks to offer new services—brokerage, mutual funds, insurance, mortgages, car loans and credit cards—either directly or indirectly from their Web

Intranets and Extranets
Internet technology that was originally developed for use over public networks to access information is now being applied within companies for remarkable gains in productivity and reductions in costs. Intranets make use of standard Web browsers to allow new types of communication, collaboration and access to data by corporate employees. Banks and financial institutions are rapidly adopting Intranets to replace aging and expensive systems in areas such as human resources management, sales automation, database record maintenance and financial reporting. Corporate Extranets apply the same technology to communication among businesses, allowing more effective interaction with vendors, suppliers and partners.

For more information on the use of Intranets, see the white paper "50 Ways to Put an Intranet to Work", available at www.usweb.com.

sites. Banks or branch offices that don't offer these services have the opportunity to co-brand offerings with specialty companies. These limited co-branded specialty offerings are often pre-packaged turnkey solutions that require limited marketing attention but provide immediate benefits. For example, many community banks offer instant approval of consumer and mortgage loans via the Internet and telephone. Rather than build this capability themselves, they outsource the processing of the data to a third-party loan processing company.

In many cases, relationships can be structured to allow institutions to participate in the revenue stream that is being created between their customers and the partner organizations. Perhaps most importantly, they allow institutions to maintain and control the relationship with their customers, while offering superior services that they might not other-wise be able to provide.

Increase Customer Retention

One of the primary reasons people change banking institutions is that they have relocated from one area to another and, as a matter of convenience, desire a bank that provides access and services in their new location. While many banks offer ATM, bank-by-mail and telephone banking services, customers often find that these services do not meet all of their needs. With the rise of Internet banking comes the ability to conduct most, if not all, of a typical customer's banking online, either via Web access or through personal finance software. The services that are offered online may be exactly the same as those available through a combination of telephone, ATM and bank-by-mail. However, customers using an online "branch" do not report feeling the same degree of isolation or the perception of being "second-class" banking customers. When a comprehensive online banking system is designed, an online "branch" gives customers the perception of actually visiting the bank, interacting with employees and conducting business, rather than trying to use the services of some physical branch from a remote location. Online branches have been shown to dramatically reduce the loss of customers due to relocation.

Extend Geographic Reach

Many banks that have significant online banking systems report that in addition to increased customer retention rates after physical relocations, they are seeing new customer growth outside their home regions. And this growth comes from new customers who have never lived in the bank's home region and will likely never visit a physical branch. Supplemented by national ATM networks, bank-by-mail services and telephone banking, some banks have begun marketing their services nationwide without having any physical locations outside their home territory. This extension of a bank's geo-graphic "draw" area to include virtually anyone on the Internet is often one of the most overlooked benefits of a well-designed, well-marketed online banking system.

Cross-Sell Services

Internet sites collect useful data in ways that are virtually impossible to collect through any other medium. Tracking

Figure 3: Banking Transaction Costs

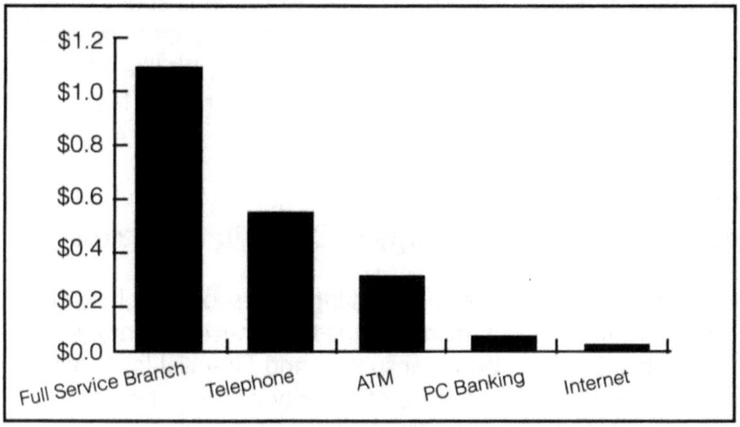

Identify Profitable Customers

As discussed earlier, the Internet allows companies to capture transaction and customer information more readily than any other financial services delivery channel. In addition to using this information to market special products and services, some organizations use customer data to determine whether or not a particular customer, or an entire customer segment, is profitable.

A comprehensive customer database facilitated by a Web site helps to identify which customers are most profitable and to target special offerings to maintain their loyalty. Not surprisingly, most banks find that Internet banking customers tend to be the most profitable ones. A 1996 Booz, Allen & Hamilton study showed that the average online banking customer uses 3.2 banking products (compared to two for the average customer). It also showed that in particular segments, Internet customers are on aver-age three times as profitable as non-Internet customers. In addition, according to FIND/SVP, the average Internet user has a household income of more than $66,700, compared to $42,000 for the average U.S. household. Similarly, Internet users tend to be better educated than non-Internet users. This combination of statistics implies that the most attractive

software allows a bank to monitor which Web pages customers (and prospective customers) view—and how long they spend viewing them. This information, when combined with customer databases, allows financial institutions to target banking products and services more effective-ly to customers. The moment an individual logs onto a Web site, a customer database can identify what products they currently have in their portfolio, which products they have inquired about and which they are most likely to purchase. Then, using personalization software, the Web site can display a unique advertisement or promotion to capture the customer's attention. This advertisement might cross-sell a product, suggest a change in account status or provide some other form of valuable information. The Internet allows banks to develop sophisticated, personalized marketing campaigns and increase cross-selling success rates.

and profitable customers are most apt to demand online banking both today and in the future.

Reduce overall costs
Internet banking reduces a bank's costs in two fundamental ways: it minimizes the cost of processing transactions and reduces the number of branches required to service an equivalent number of customers. According to the American Banking Association, the average cost for a full-service branch transaction is roughly $1.07. Since an Internet banking transaction links directly to the back-end processing system, an Internet transaction costs roughly $0.01.

The savings associated with Internet transactions are even greater due to the small incremental cost of servicing additional Internet customers compared to the large cost of opening a new branch. It's important to note, however, that many banks don't anticipate realizing the true benefit of transaction cost reduction until a larger percentage of their customers use the Internet as a primary delivery channel.

Identify New Fee Services
Some banks use Internet banking as an opportunity to generate additional revenue from customers. Typically, fees range from free to $14.95 per month, depending on the level of service, number of transactions and type of account. It is still unclear whether or not banks will continue to be able to generate fee revenue from online banking. It's likely that Internet banking pricing will mimic ATM pricing. The cost to consumers will drop down to free but may eventually rise to a per transaction fee. However, given the new types of products and services an institution is able to offer and the increased ability to cross-sell and target profitable markets, it's likely that the Internet will generate increased fee revenue for most banks.

The Biggest Obstacle to Success

While there's little question that Internet banking offers substantive advantages to both consumers and financial institutions, the issue of security is often cited as a major barrier to widespread consumer adoption. While many of the actual security issues today have been addressed with recent technical advances, financial institutions may find that consumers still perceive a bigger problem than there really is. A comprehensive online banking solution must address real security issues as well as the psychological or perceived security fears of consumers.

The past two years have seen remarkable advances in the ability to provide cost-effective, highly secure transactions over public networks. With the wide-spread adoption of secure firewalls, 128-bit encryption schemes and digital certificates that accurately identify authorized users, online purchases and transactions have skyrocketed. In fact, many experts now say that a well-designed online banking system is less vulnerable to fraud and attack than a physical bank location or an ATM. While no system could ever be called "perfectly secure" any more than a branch office could be, the cost to the institution of providing extremely high levels of security is much lower for a complete online system than is typically spent to "secure" a single physical location.

With the technical issues of security properly addressed, the institution must be able to reassure consumers that security and privacy problems have been solved. Financial institutions are turning their focus to educating their customer base about the types of security put in place on the Internet to protect their financial information and what to look for to ensure that their transactions are secure. Additionally, banks must be sensitive to issues of privacy and provide customers with written assurance that their personal information will not be exploited.

The Four Phases of Internet Banking

A bank typically considers four phases in developing an Internet presence:

- Phase One: Marketing and Promotion
- Phase Two: "Light" Interactivity
- Phase Three: Full Transactions and Services
- Phase Four: Strategic Usage

While most banks migrate from phase to phase in the order given, some skip Phases One and Two altogether and move directly to Phase Three: Transactions and Services.

Table 1 highlights some of the key distinctions of the four phases.

Table 1

	***Phase One:* Marketing and Promotion**	***Phase Two:* Light Interactivity**	***Phase Three:* Full Transactions and Services**	***Phase Four:* Strategic Usage**
Focus	*Marketing Web site*	*Customer acquisition*	*Banking functionality*	*Strategic change*
Primary Services	• Published information on bank services • Branch/ ATM map • Customer service e-mail	• Loan calculators • Credit card applications • Savings, checking account applications • Financial Planning articles, advice	• Account look-up balances, transfers • Bill payment • Car loans credit cards, mortgages • Statement review • Cleared check presentment	• Sophisticated cross-selling of new services • Customer profitability analysis • Bill presentment & payments
Primary Benefit	• Provide information to customers and prospective customers	• Reducing paperwork • Low-cost ways to attract and impress customers	• Retention of existing customers • Attracting high-value customers • Reduction in service costs	• Increased service offerings • New revenue opportunities • Increased margins

The resources required to pursue these phases are highly dependent on a bank's back-end processing system and technology infrastructure. There is not necessarily a correlation between bank size and the phase in which the bank is currently operating.

Phase One: Marketing
Launching a Phase One presence involves very little risk. Generally, banks consider Phase One to be a learning experience. They use existing marketing and promotional material, adapt it for the Internet and develop a Web site. Phase One requires mini-mal investment, limited maintenance and limited resources. Often only one or two employees are involved in creating and maintaining the Web site, thereby reducing the burden on the organization. In fact, many banks outsource the creation and management of the Web site to a qualified specialist or service firm. Engaging in Phase One provides the opportunity to learn about the Internet and to signal to customers the importance of the Internet while promoting traditional services to a wide audience.

Phase Two: Light Interactivity
Phase Two involves light interactivity. In this phase, banks enhance their public Web sites in an effort to provide more value to existing and prospective customers. Examples of services or functions offered in this phase include online loan calculators to help customers determine their payment schedules, credit card and loan applications, and stock or mutual fund information. Banks have experimented with regular articles about financial planning, as well as explanations of financial terms and products. These services are designed to support existing customers and keep them coming back to the Web site on a regular basis.
In addition, well-designed free services on the Internet give banks an opportunity to promote their services to new customers who visit the Web site to make use of these free tools. What distinguishes Phase Two from Phase Three is that in this phase there's no link from a bank's back-end processing system to its Internet Web site. A Phase Two deployment requires a larger investment than a simple marketing site but does not require the larger systems

Figure 4: Types of Online Consumer Banking Sites

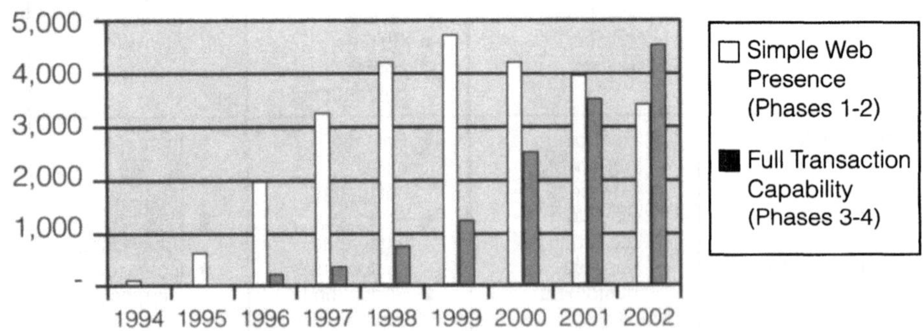

Source: Online Banking Report, January 1998; modified

integration investment of a full online banking system.

Phase Three: Transactions and Services
Phase Three offers true online banking over the Internet, including balance transfers, account statement access, bill payment services, account applications, check image retrieval and customer service mechanisms.

As outlined in this paper, the benefits to a Phase Three deployment can be considerable, and there are many motivations for pursuing this phase. Some banks migrate to Phase Three to retain key customers; others to acquire new ones. Some banks consider it a defensive move—they fear they will lose their best customers if they don't offer online banking. Others are looking to reduce over-head costs either in physical branch operations or customer support resources. Still others consider Phase Three an opportunity to generate fee income and to offer services

Banking Standards
When a financial institution launches a Phase Three or Phase Four Web site, it links its back-end processing system to an Internet server. Depending on the type of back-end processing system employed, this can be a substantial challenge. Linking legacy mainframe systems to a Web interface in a secure, fault-tolerant environment is the primary obstacle associated with the migration from Phase Two to Phase Three.

Two standards have emerged to help banks link an Internet server to a legacy, back-end processing system. The first is Open Financial Exchange (OFX), which combines Intuit's Open Exchange, Microsoft's Open Financial Connectivity (OFC) and CheckFree's extensions of these standards. OFX is a technical framework that provides a comprehensive approach to financial data exchange and is increasingly being adopted by back-end processors to simplify the process for conducting transactions online. One of the primary benefits of OFX is that the standard transaction language is supported by both Quicken and Money, the leading personal financial management software packages. At the same time, OFX permits access to the back-end systems via a Web site, allowing both types of users to get at the same data without the bank having to invest in two completely different connectivity solutions. The second online banking standard is Integrion's Gold. Gold focuses on creating a framework for electronic messaging between the Internet and a bank's processing system; it also includes tools to help a bank bring new products and services online.

To most banks, it doesn't really matter who wins the OFX vs. Gold battle. What's more important is that there are frameworks that will jump-start the process of offering comprehensive transactions over the Internet—no matter how complex the back-end system is. It is likely that both standards will continue to co-exist, or they may eventually be merged into one.

they otherwise wouldn't be able to offer, such as credit cards, mortgages and mutual funds.

Phase Three deployments are considerably more complex than those of Phase One and Two, because of the systems integration requirements in connecting a public Web site to existing transaction systems. In addition, Phase One and Two deployments do not have the security concerns of Phase Three, since actual account and customer information is not avail-able. Fortunately, due to standard protocols and security products now available, a Phase Three banking deployment costs a fraction of what it did even a few years ago, while offering all of the benefits described in earlier sections of this paper.

Phase Four: Strategic Change

In Phase Four, banks revolutionize transaction processing and customer interaction. They replace investments in "bricks and mortar" with investments in advanced Internet services and Web technology. They develop sophisticated customer databases to enable advanced customer segmentation to increase cross-sell opportunities and customer retention. They build efficient targeted marketing campaigns that boast success rates far greater than direct mail campaigns. Web-enabled decision support systems within the organization improve the creation of both targeted online and direct mail campaigns by giving bank marketing departments access to critical information and the ability to query and run simulations against these sophisticated customer databases. These advances allow banks to experiment with new ways of exceeding their customers' expectations.

Leaders in this area are launching bill presentment options and electronic commerce solutions, and they are experimenting with Smart Cards. They're creating intelligent agents that watch for particular trends in a customer's banking habits, or changes in financial or family situation, and then make recommendations for how to restructure investments or take advantage of new services appropriate for that particular customer's needs. They're using the inherent capabilities of the Internet and sophisticated databases for servicing their customers in ways that aren't possible with the traditional banking model.

For most banks, Phase Four is a vision of the future. However, it will also be the battleground of competition among financial services companies in the near future. The institutions that make the best use of the sophisticated data they have can service their customers better while improving their operating margins, and they will be the winners in the battle for customer assets. Investments in tracking, database and decision support systems to manage, analyze and use this data are wise investments for financial services firms today.

Developing an Internet Banking Offering

Regardless of which phase of online banking is appropriate for a specific financial institution, the development and deployment of a solution involves four major tasks:

- Step One: Strategy Development
- Step Two: Analysis and Design
- Step Three: Technology Development
- Step Four: Implementation

Step One: Strategy Development

In formulating strategy, it's essential to clearly articulate the objectives of the online presence. Is the institution attempting to increase margins, reduce costs, attract new customers, broaden its geographic presence or increase customer retention? Is the institution seeking to expand its own offerings to customers or partner with appropriate organizations to offer these services? Does the bank know who its most profitable customers are, and what those customers need and expect from an online banking solution?

It is also essential at this stage to carefully analyze the strategic landscape of the marketplace. Who are the bank's real competitors? What are their offerings?

A good strategy clearly defines the business reasons for developing a Web site and helps choose which phase of Internet banking to consider.

Step Two: Analysis and Design

With the strategic groundwork in place, a bank is prepared to identify technology requirements and plan the system architecture that supports those requirements. It is at this phase that the bank (or its consultants) will need a detailed understanding of the competing hardware platforms, connectivity solutions and financial transaction protocols used in the marketplace today. If the deployment is a Phase Three offering, much of the time and costs at this phase will be in the areas of connectivity with existing transaction systems, and in the careful planning of security and customer authentication systems.

A critical step often overlooked at this stage is the plan for how the system will be rolled out to existing customers and marketed to prospective customers. Issues such as the overall graphic look of the site, the online support and training information that will be necessary, and the marketing plan for attracting users to the system must all be contained in the roll-out plans.

At the end of Step Two, the project team will have detailed technical documents that delineate the overall system architecture and the functional specifications of the system to be deployed, as well as a comprehensive plan for how the system will be announced and introduced to the market.

Step Three: Technology Development

Once the functional specifications and system architecture plans are finalized, the system is ready to be built. At this stage, systems and hardware are acquired and deployed. This may take place either on site at the bank, at its transaction processing vendor's location or at a third-party hosting center, depending upon the strategy laid out in Step Two. The appropriate systems integration code will be written, tested and deployed to ensure that the relevant systems are able to communicate with each other effectively. The security and authentication systems will be deployed and thoroughly tested. At the conclusion of this step, partially or fully functional versions of the major components of the sys-tem will allow the solicitation of feedback and the identification of problems before the institution moves on to Step Four.

Step Four: Implementation

In the final step, the team assembles and tests the full production system from the technology components acquired or built in Step Three. Employees are trained in the use and support of the system. Also, the bank begins the execution of the roll-out strategy developed in Step Two to help customers understand the new offering. Finally, the bank will adopt systems for measuring the success of the project and develop a plan for maintaining and updating its Web site.

Web sites are rarely static. Once Step Four is completed, it's likely that the bank will return to Step One to update its Web site or to add functionality.

Five Tips for Establishing an Internet Banking Presence

Plan carefully before you design

It is essential that an online banking presence support the overall business strategy of a financial institution. Before any code is written or any systems purchased, be sure that you know what your overall business goals are and how your online presence will help you attain those goals.

Utilize accepted, open standards

The sidebar on page 12 talks about the two most prevalent standards in online banking: OFX and Integrion Gold. The selection of a particular standard will depend upon the back-end processing systems you use and the functionality you wish to offer your customers. Adhering to one or more of the major financial interchange standards ensures that your offering will support users of various software pack-ages, reducing your development and maintenance costs. In addition, a system designed to support security standards such as SSL and SET ensures that your system takes advantage of the security features built in to the current generation of Web browsers.

Make it scaleable

Be sure the Web site you build today can be scaled up to meet your future needs, including the range of services offered, the number of customers able to be served and geographic areas in which you may one day operate. A sys-tem built on technology that does not scale well will represent a wasted investment as your needs change and your system grows.

Conclusion

Online banking has finally reached the point at which it is being accepted by consumers and is financially successful for the offering institutions. In fact, banks are finding that a comprehensive online banking strategy is essential for success in the increasingly competitive financial services market. Due to technology advances and the rapid growth of the Internet, an online banking solution can be designed today that is more effective and less costly than the proprietary systems of only a few years ago. Banking transactions can now be initiated and monitored via standard Web browser software or the major financial planning software packages. The software offers a range of features for both the casual user and the sophisticated banking customer. In planning an online banking presence, an institution needs to consider various options, ranging from a simple marketing presence on the Internet to an advanced online banking system that revolutionizes the way in which the bank operates and the consumer conducts financial transactions. Regardless of the type of offering a bank decides to pursue, it is essential that the bank undertake a comprehensive analysis and planning process up front to ensure the success of the project.

Keep your brand
One of your greatest assets is the relationship you maintain with your customers. In developing an online banking strategy, be sure to keep ownership of this relationship and look for opportunities to enhance it. If you pursue home banking options that operate through financial management software packages, carefully examine the options for keeping your brand prominently in front of your customers. If you do not demand and take advantage of these features, there is a danger that your customers will shift their loyal-ties from you to the software maker.

Work with capable, experienced consultants
Very few banks have the technical capabilities to design, plan and deploy an online banking solution on their own. Find a professional services firm that offers a complete range of services, including strategy consulting, systems design, integration and deployment. Even if your initial goals are for a simple Web site, look for a consultant that has the technical knowledge and experience necessary to design an advanced online banking solution. This will ensure that the relationship you build with your consultant is maintained as your system gradually evolves in sophistication. Also seek independent consultants, rather than those affiliated with a particular software product. This will ensure that you receive unbiased advice. Your selection of hardware, software and connectivity products should be based on your own specific needs and situation.

About USWeb

USWeb Corporation, headquartered in Santa Clara, California, is a leading international professional services firm dedicated to helping businesses solve their mission-critical objectives through Web-based technology.

As a single source for Internet and Intranet solutions, the company offers a comprehensive set of services including needs analysis, architecture planning, strategic planning, Internet/Intranet/Extranet consulting, design, development, deployment, access, hosting, site marketing and educational services.

USWeb delivers custom solutions to streamline business processes such as marketing, engineering, distribution and operations, HR management and recruitment, sales force process management, customer-service and support, e-commerce and workgroup collaboration.

USWeb has built its business on providing our clients with successful custom Internet and Intranet solutions.
We can provide the expertise you might not have in-house to save you time and money and get your Internet/Intranet solution up and running more quickly.

To find out how USWeb can deliver an effective custom Intranet and Internet solution for your company, call 1-888-USWEB-411, or visit our Web site at www.usweb.com to locate the USWeb office nearest you.

BankAway! – Internet Banking – Issues & Challenges

Title: *BankAway! – Internet Banking – Issues & Challenges*

Author: *Infosys*

Abstract: *Internet Banking is a new concept in most regions and as expected fraught with numerous challenges both on the technology and business side. This document looks at some of the key issues and challenges banks are facing and how they are being addressed.*

INTRODUCTION

Until recently the Internet was thought to be a well-suited medium for information exchange and marketing purposes. However, the likes of DELL Computers, CISCO, CITIBANK and amazon.com have made people sit up and realize the potential of this medium as a channel for transacting business.

It is no great surprise that recently secure Internet Commerce has been the focal point at many businesses across the world and has created much of the excitement in the technology world. An exponentially growing population of over 150 million, young, educated, high earning, technology savvy individuals is something no business can afford to ignore, banking is no exception to this.

With a large number banks offering ATM and Phone banking services since early nineties, the idea of virtual banking was never new but the advent of the Internet has probably given this idea the greatest momentum.

Internet banking offers the lowest transaction costs among all the electronic delivery channels Œ ATMs, Telephone and PC. A study conducted by Booz-Allen and Hamilton stated that the cost of an average transaction on the Internet could be as low as, 13 cents, compared to $1.07 through the branch, 54 cents through the

Figure 1: Worldwide Internet Commerce

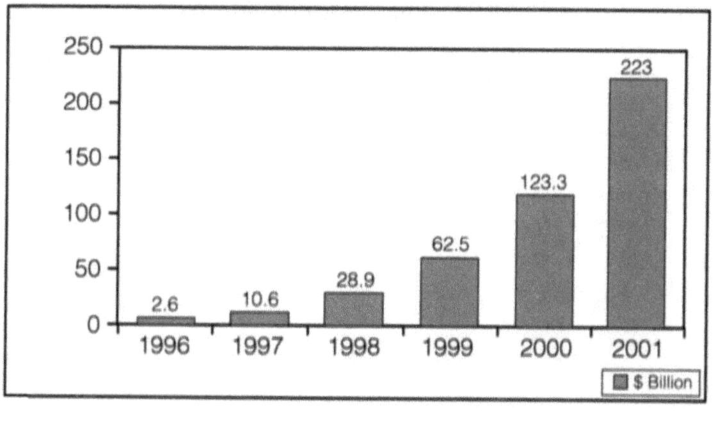

telephone & 27 cents through the ATM, for a similar transaction.
In addition to this, the cost of building brick & mortar branches has been escalating whereas the cost of communication and computing equipment has been constantly on the decline.

Banks have taken cognizance of this fact and there are :

- Over 2500 banks who have created a web presence
- More than 500 of these are offering account access services.

For those who thought all this is an America only phenomenon can take cue from the fact some of the most advanced services i.e. Online Loans, Portfolio Management, Smart Card Loading etc. are known more outside of America than within.

The reasons for this are not too difficult to guess:

- Banks in the North American continent are largely on mainframe-based systems and implementing the entire set of value added delivery channel services for customers would require massive revamp and upgrade of these systems.
  In the Asia and parts of Europe however, most banks are already on Client/Server and hence, implementation of web based technologies is already underway.
- The interactivity and user friendliness of Internet technologies have made it possible for businesses to give each customer a unique experience based on his profile. Thus, banks, who operate in geographies where the heterogeneity of customer profiles is high, prefer to use the Internet where the experience for each customer is both, unique and consistent.
- Over 90% of the world population lives outside of North America.
  A penetration of the Internet as low as 1% would translate into several million users especially in places like South Asia which have one of the largest population densities. Thus, economies of scale are much easily possible in such regions even with low Internet penetration.

The usage of on-line financial services seen from the chart has been on the rise. Also, more and more banks are building ‚Internet Branches™. It is expected that by the end of Year 2000 over 3000 banks will offer complete set of Internet Banking services and it is fair to assume that a large chunk of these will be outside North America.

Figure 2: Usage of On-line Financial Services

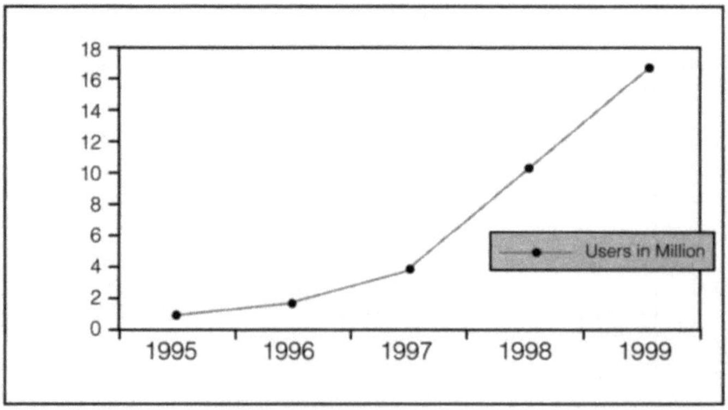

KEY BENEFITS

The main driver for the trends above no doubt is the significant reach of the Internet. However, the Internet does offer other benefits to the customers as well as the banks.

THE BENEFITS THAT THE CUSTOMERS CAN EXPECT ARE:

GREATER CONVENIENCE
All the banking transactions can be performed from the comfort of the home or office or any other place the customer wishes to. This is especially useful for the customers who are always on the move, they do not have to waste time trudging to the branch for every banking need.

BETTER FUNDS MANAGEMENT
The customers can download their history of different accounts and do ,"What – if" if analysis on their own PC before affecting any transaction. This way a savvier customer can better manage his funds and payments. Also, since most banks are offering online transaction capabilities, the customers can actually wait till the last minute before moving funds thus, optimizing use of funds.

THE BENEFITS THAT THE BANK GETS ARE:

REDUCED BRANCH LOAD
It has been found that the load of routine transactions at the branch e.g. Œ generating statements, balance enquiry, cheque book request etc. increases proportionately with the increase in number of customers. This poses a challenge to the bank in terms of maintaining quality & consistency of service at the branch.
The use of the Internet helps the bank reduce the load of routine banking transactions at the physical branch. This is useful especially in branches catering to a large number of customers. Also, due to this, the bank can address a greater customer base while maintaining the same number of branches and staff strength i.e. in a way, the banks return on assets employed increases significantly.

ATTRACTING FUTURE CUSTOMERS
Most market research has shown that a typical Internet user is a young, educated, technology savvy person. These people may not have huge bank balances today but definitely a large number of them might have it in the near future.
Thus, the Internet is a critical medium, which offers the bank a strategic advantage in sustaining future growth and profitability.

LEVERAGE THE POWER OF VISUAL COMPUTING
The PC can support much more interactivity in the form of video, sound, multi-media etc. than that offered through other available channels e.g. the phone or the ATM. This interactivity can be used to personalize the banking experience for each customer and also, to deliver a host of information in the form of tickers / animated gifs etc. to his desktop.

BUILD THE BANK'S BRAND
By use of the Internet, banks can build and reinforce their brand worldwide, at a very nominal cost, owing to the global reach of

the Internet. This benefit especially accrues to the early movers / innovators.

ISSUES & CHALLENGES

Internet Banking is a new concept in most regions and as expected fraught with numerous challenges both on the technology and business side. This section looks at some of these key challenges facing banks and how they are being addressed.

TECHNOLOGY CHALLENGES

SECURITY OF TRANSACTION INFORMATION

The Internet is a public network with over 150 million users and more than 25 million host computers connected. A financial transaction that is sent on this network could traverse the world several times before it reaches the final destination. This poses a threat for all banks as, the message transmitted can be trapped or altered at numerous locations. Security standards such as SSL & the Public Key Infrastructure have emerged as the defaults to address this issue. PKI technology involves asymmetric encryption algorithms. These allow the bank and the customers to encrypt and safely transmit transaction information over the web. Thus, even if the messages are trapped the risk of their being interpreted is insignificantly low.

SECURITY OF BANKING ENTERPRISE DATA

For enabling either information access or transactions, banks have to put up a web site. There is data to suggest that most hacking activity that happens is targeted at breaking into web-site security. Typically hackers will break into a web-site first and then try to get access to internal systems from there. This is perhaps the most serious security threat for all banks. Most banks have started looking at setting up secure sites and network architecture which involved not just hardware & software tools but also, the processes which need to be followed in order that the web-site is secure from all such threats. No security solution can offer a panacea for all known threats but they do fortify a large part of the risk involved.

" *There still exist impediments to Internet Commerce even today but, they seem to be more of business and organizational issues than technical ones* "

INTEGRATION WITH MULTIPLE AND DISPARATE SYSTEMS

In order that the customers can access their information over the web, the bank not only has to have a self-service application running on its web-site but also has to integrate with the existing core banking system(s). This interface can be in two modes, off-line or store & forward and online or real time.

Most banks that looked at Internet banking two years ago implemented the offline mode as, they felt real time transactions were less secure and also, the transaction loads did not justify having a real time interface. Also, they preferred this mode because the legacy systems existing in their environment. These systems though have mutated and have evolved from the 1970s era but they still retain a large part of their original architecture. Interfacing with legacy systems has always been a key issue e.g. customer information files that are designed to provide relational access to customer and his relationship with the bank are fairly recent and are not ubiquitous. Hence, offering accounts access on these systems too is a big challenge.

For banks who have advanced client/ server systems and who have implemented ATM and phone banking earlier, the key challenge is to give a single, unified and real-time view across all the delivery channels. This is necessitated for customer service reasons and also owing to concerns related to fraud. E.g. if the databases from which the tellers operate and from which the ATM withdrawal happens, are different, there is a possibility that people can take advantage of this and overdraw their accounts.

BUSINESS CHALLENGES

ACCEPTANCE & GROWTH
The transaction costs of the Internet have been stated to be the lowest of all channels. However, the total volume of benefit to the bank will become significant only when a large number of transactions are routed over the web than over other traditional channels.
This requires that the bank have a large chunk of Internet savvy customers who can exploit this medium to its full. The number of users having access to the Internet is increasing by each passing hour but, a large part of the population is yet to discover the entire set of possibilities on this medium i.e. the number of savvy users in most places is still low.
The Internet and the interactivity have caught the attention of many a layman to the extent that in many places the need for the Internet itself is driving the sales of PCs up. It is a matter of time before a large chunk of the user population becomes ,NET SAVVY.'.

" *In the first two years after the launch of a new application, people tend to overestimate its use. In the next five years they underestimate it*"- *Bill Gates*

PUBLICITY OF INTERNET BANKING SERVICES
At times awareness levels about Internet Banking remain low even after a bank has launched the service. This can be attributed to that insufficient resources are allocated to promotion of the service to customer. As required for any other product launch, banks should have a marketing and promotion planned clearly chalked out. Also, banks can look at branding this service and creating internal organization structure and processes to manage this brand. This will not only help in awareness creation but also ensure that the customers start exploiting this medium.

CREATING A COMPELLING INTERACTION
Most banks are content with launching just account access services. These services though are useful but are not value adding to the extent that the customers starts using the Internet Banking service on a regular basis. Thus, bank's needs to keep introducing more and more value added services. Ultimately, the banks web-site should become a portal for all financial needs of its customers. The services being launched currently are useful only for the retail customers. These customers typically have lower loyalties and also lower cost of switching to other banks. Thus, any bank

Figure 3: Returns on interactivity

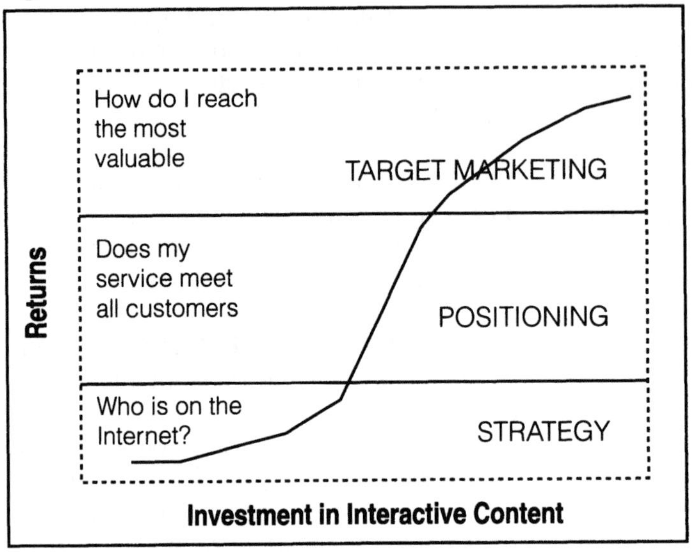

can easily lose these customers to other banks or non-banking competitors who start offering these services as, it is not difficult to address the needs of retail customers.

Thus, the banks need to look at all customer segments e.g. corporations or businesses and create value for them too. In fact, in some regions it is the businesses that have ready access to the Internet. Thus, in such a scenario, creating a compelling interaction for them might make better business sense.

PERSONALIZING THE BANKING EXPERIENCE

With the onset of electronic delivery channels including the Internet, the one-to-one personal interaction with the customers is likely to reduce. Personalized attention and service has been the corner stone for many a banks for decades and thus, losing this might ultimately mean losing customers.

The bank needs to look at ways of personalizing the experience of each user even on the Internet. This implies building a lot of intelligence in the software systems.

In fact some banks are looking at deploying Casta-nets. A typical casta-net has two modules. The first one rests on the customers' PC and the second run at the bank.

The bank can keep pushing information on various topics (known as channels) e.g. financial news, currency rates, interest rates, securities, mutual funds etc. and the customers can browse through the information based on the channels they have subscribed to. The concept is similar to that of television broadcasting where people just have the receiving equipment in their homes and they can view the programs depending on the channels they have subscribed to / paid for.

SUMMARY

Internet Banking does raise quite a few issues, some which may be quite challenging for banks. The biggest amongst these perhaps is that the traditional business paradigms are demolishing much in the same way as a brick & mortar ,STRUCTURE would. Increasingly banks are being forced to consider broadening their focus beyond traditional paradigms. In this regard - the business focus, operating scale and existing systems of the bank might require a close look. However, a little effort can create innumerable new strategies, relationships, customers and ,STRUCTURES.

Bankaway

The e-commerce platform

BankAway is the powerful E-Commerce platform that enables your bank to offer diverse services like retail banking, corporate banking, investment, third-party payments and shopping over the Internet with complete customer orientation.

COMPREHENSIVE ON-LINE BANKING
With BankAway, not only retail customers but your corporate customers as well can bank on-line.
Using BankAway, customers can open an account with your bank online, and begin accessing them.
BankAway also, empowers your bank to provide an integrated financial service offering to your customers Œ the one click access to retail banking, corporate banking, investments, credit card payments, on-line shopping and utility payments.
Thus, BankAway enables your bank to address a large segment of the customers.

PAYMENT ENGINE FOR INTERNET SHOPPING
The unique payments engine enables Internet Shopping malls to link up with the bank to enable payments for customer purchases through their accounts i.e. even those who do not have debit/credit cards can enjoy Internet shopping through your bank.
This powerful feature ensures that your bank plays a vital role as a HUB in E-commerce transactions between merchants and consumers.

TRUE RELATIONSHIP BANKING
Understanding the banks' need to respond to the customers requirements quickly and flexibly, BankAway is designed to enable the bank to use customer information to provide each customer with a personalized experience.

MULTI-LAYER SECURITY
Security is a key requirement for financial transactions over an open network. BankAway addresses security at multiple levels Œ transmission level, Web-server level and application level. BankAway matches up to the most secure financial solutions in the world with open standards like Secure Sockets Layer (SSL) and asymmetric key encryption.

SECURITY CONSULTING SERVICES
Along with powerful security features in BankAway, Infosys Technologies also offers security-consulting services to the bank. Experts from Infosys work pro-actively to assist the bank in setting up the secure environment necessary for implementing on-line banking. This service is a unique part of the Infosys Offering.

MULTIPLE PLATFORMS
Based on open standards, BankAway enables the bank to easily support transaction processing on the Internet. The bank has the choice of either operating on

Windows NT or Solaris platforms. Also, the robust architecture ensures that it is extremely scalable and can intelligently handle varying load of customer transactions.

INTERFACE TO EXTERNAL SYSTEMS
BankAway has capabilities to interface real-time with diverse systems and networks e.g. Core banking systems, ERP systems, Bill payment service providers, Payment networks and other value added service providers / networks, ensuring the banks lead in the E-Commerce market-space. All this and much more guarantee that BankAway can help the bank comprehensively meet the challenges of the new era.

Electronic Banking in Malaysia: A Note on Evolution of Services and Consumer Reactions

Title: *Electronic Banking in Malaysia*
Author: *Balachandher Krishnan Guru, Santha Vaithilingam, Norhazlin Ismail, Rajendra Prasad*
Abstract: *Developments in information technology and telecommunications have set in motion an electronic revolution in the Malaysian banking sector. This in turn has resulted in new delivery channels for banking products and services such as the automated teller machines (ATM's), telebanking and PC-banking. In this context, the purpose of this study is to examine the evolution of electronic banking in Malaysia, analyze the various electronic delivery channels utilized by local banks and to assess the consumers' reactions to these delivery channels.*

Biography: see end article

Introduction

Technological developments particularly in the area of Telecommunications and Information Technology are revolutionizing the way business is done. Electronic commerce is now thought to hold the promise of a new commercial revolution by offering an inexpensive and direct way to exchange information and to sell or buy products and services. This revolution in the market place has set in motion a revolution in the banking sector for the provision of a payment system that is compatible with the demands of the electronic marketplace.

In line with global trends, banking business in Malaysia too has been undergoing tremendous changes since achieving independence in 1957. The first step in the evolutionary process was the gradual deregulation of the financial sector, which commenced in the 1970's. Then, in the 1980's the introduction of Automated Teller Machines (ATM's) can be considered as the first and most visible piece of evidence of the emerging electronic banking in Malaysia. This was then followed by the introduction of telebanking and later on PC-banking in the 1990's. The next imminent step in this evolutionary process inevitably appears to be Internet–banking. However, there are several issues including the lack of adequate legal framework and security of electronic transactions, which tend to hamper the continued progress of this evolutionary process.

The objective of this paper is fourfold. First, it presents a discussion of the progressive developments of electronic banking in Malaysia. Second, an analysis is made of the types of electronic delivery channels utilized by local commercial banks. In this context, the Malaysian commercial banks' use of the World Wide Web (WWW) is also assessed. Third, some pertinent issues for the successful implementation of electronic banking are discussed. Fourth, the paper presents the findings of a questionnaire survey carried out to evaluate the Malaysian customers' perception of electronic banking services. Finally, the paper is concluded with a summary of the findings and a discussion of the limitations of the study.

The Electronic Revolution in the Malaysian Banking Sector

The electronic revolution in banking basically centers on changes in the distribution channels of financial institutions. The basis for the emergence of the modern electronic distribution channels is the result of the evolution of the concept of money. In the days of barter trade, the ability to pay for goods and services was reflected in the physical existence of the goods, which could be used for exchange. Then, came hard cash in the form of coins made out of precious metals. This was then followed by the advent of fiduciary money in the form of modern coins and paper notes. Today, an individual's ability to pay for goods and services is simply reflected in the accounting records of his or her bank. Thus, it is important to appreciate at the outset that money as it is defined today is just simply information, which can be electronically transmitted to facilitate economic transactions. It is this new definition of money, which has resulted in the electronic revolution of financial institutions.

The first giant step in this electronic revolution was the computerization of financial institutions. According to Pang (1995), computerization in the Malaysian banking industry got off to a slow start and only picked up momentum in the 1970's. The increasing volume of banking transactions was the inevitable motivator for the introduction of computers in Malaysian commercial banks. Then, by linking up technological developments in telecommunications and Information Technology, real-time on-line electronic funds transfer came into existence. A large part of the electronic funds transfer process takes place within the banking premises and thus may be invisible to the lay person.

Today, Malaysian commercial banks have the privilege of various delivery channels for their products and services. This includes the brick and mortar branch office networks, automated teller machines (ATM's), automated self banking channels such as the BSN Commercial Bank's Electronic Banking Center (EBC) and Phileo Allied Bank's virtual kiosks, telebanking via the telecommunication channel and PC-banking.

Figure 1: The growth of ATM's in Malaysia

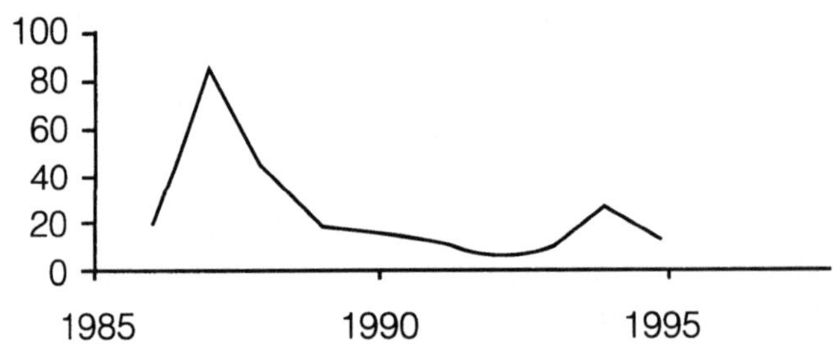

Source: BNM Annual Reports (1985-1997)

Automated Teller Machines

The most visible form of electronic funds transfer in relation to the Malaysian banking sector, was the introduction of Automated Teller Machines (ATM's) in 1981. The ATM's released banks from the constraints of time and geographical location. They presented banks with a more economical substitute for brick and mortar branches. The revolution brought about by this technology were longer banking hours which extended beyond office hours.

Another immediate impact of the introduction of ATM's was that bank staff could be relieved of some mundane functions like processing withdrawals and fund transfer over the counter. Today, the ATM's in Malaysia can be used for balance enquiry, cash withdrawal, transfer of funds between checking, savings and credit card accounts, bill payments, making payments to application for initial public offerings in the Kuala Lumpur Stock Exchange and for making cash and check deposits.

However, Malaysian bank customers had not been very confident of making cash deposits via the ATM since the cash is only physically deposited into the bank but only credited into his or her account usually on the following working day. Furthermore, the bank officer has the final say on the amount deposited in the event of any dispute.

As at the time of writing, local banks such as BSN Commercial Bank, Maybank, Phileo Allied Bank, Hong Kong Bank and the Standard Chartered Bank in Malaysia have introduced cash deposit machines (CDM's) that can accept cash which is immediately verified and credited to the customers account. This technology was however implemented by Posbank in Singapore as early as 1997 where the ATM actually sorts and counts the money and asks the customer to confirm the amount. If the amount tallies the customers account is immediately credited but however, if the amount does not tally, then the amount is automatically returned to the customer. It is this capability of the machine to perform like human teller, which has succeeded in weaning customers away from the counters and to the automated teller machines.

In the initial stages of development of ATM's, financial institutions installed ATM's in practically every strategic location with the primary objective of securing competitive advantage.

This is clearly indicated in Figure 2, by the rapid growth rate of ATM's in the early years. However, the result was a large number of ATM's in these strategic locations with resultant returns that did not commensurate with the heavy installation costs. Thus, in view of the high costs involved in ATM operations and the duplication of ATM services at many off-branch premises, the financial institutions agreed to pool their resources together to establish ATM network switches. By linking the respective ATM systems of these financial institutions through the switch, cardholders of member institutions of the consortium would be able to carry out transactions at the terminal of any of the other participating financial institutions.

Today, almost all the domestic commercial banks in Malaysia are members of the Malaysian Electronic Payment System (MEPS) Sdn Bhd. and the member banks' customers can have access to their accounts via any ATM belonging to the MEPS network. This may be one of the reasons for the slower growth rate of ATM's in the more recent years.

The impact of the above development was that, commercial banks with geographically well spread ATM networks like the Malayan Banking Berhad, one of the largest commercial banks in Malaysia may tend to lose any competitive edge associated with having the largest number of ATM's in the country. However, if a bank customer was to use the ATM facilities of another bank for his transactional needs, a minimum access fee of RM1 will be charged to the customer. Thus, those banks with large number of ATM's can now profitably use these machines as a source of revenue.

So what are the future prospects for these ATM's? Philip (1996) sees ATM's evolving into virtual branches providing a broader range of remote transactions where customers could interact with bank personnel through video conferencing.

Telebanking

Telebanking can be considered as a form of remote or virtual banking which is essentially the delivery of branch financial services via telecommunication devices where the bank customers can perform retail banking transactions by dialing a touch-tone telephone or mobile communication unit, which is connected to an automated system of the bank by utilizing Automated Voice Response(AVR) technology. Telebanking has been in Malaysia since the early 1990's. The telebanking service provides yet another alternative to almost all of the functions available on the Automated Teller Machines except withdrawal and deposit of cash. The facilities available include checking account balance, funds transfer between current, savings and credit card accounts and bill payments. Some of the local commercial banks like BSN Commercial also allow stock exchange transactions to be performed via the telebanking facility.

According to Leow (1999), telebanking has numerous benefits for both customers and banks. As far as the customers are concerned, it provides increased convenience, expanded access and significant time saving. On the other hand, from the banks' perspective, the costs of delivering telephone-based services are substantially lower than those of branch based services. In addition, there are about four million users of fixed line telephone services, which would certainly guarantee the critical mass criteria for telebanking services. Despite all these advantages, according to an article in the Malaysian Central Bank, Bank Negara Malaysia's (BNM) annual report only eight commercial banks offered telebanking services as at end of 1995. As at April 1999, only nine out of the twenty-three local commercial banks were known to offer telebanking services. This indicates that telebanking is currently not a major delivery channel for

Malaysian commercial banks' products and services.

The poor customer response to telebanking may be due to the fact that one of the most important transactions, which almost all bank customers perform on a regular basis namely cash withdrawal is not possible via telebanking. Another, reason for the lack of success of this delivery channel may simply be poor marketing of this product on the part of the local commercial banks. It may also be due to the lack of customer confidence in online transactions, which are not immediately verified with black and white statements or receipts. This may however be contrary to the developments in Europe where it has been reported in Leow (1999), that 95% of European banks are considering telebanking services to be offered by the turn of the century.

PC-Banking

The increasing awareness of the importance of computer literacy has resulted in increasing use of personal computers in Malaysian households. This would certainly support the growth of PC-banking which would virtually establish a bank branch in the customers home and offer 24 hour service seven days a week.

As at end of 1995, according to BNM(1995), ten local Malaysian commercial banks are already offering PC-banking or home banking services. At present, customers can use their personal computers at home or at their office to access their accounts for transactions by subscribing to and dialing into the banks' Intranet proprietary software system by use of password.

Though, PC-banking has been around in Malaysia for a while, especially over the latter half of the 1990's the number of customers who use this service is still relatively small. One of the reasons for this may have been the small number of customers who actually have access to personal computer. However, this situation is fast changing with the Malaysian government's emphasis and commitment in its efforts towards transforming Malaysia from a production economy to a knowledge-based economy. Homes with computers are fast increasing and as at 1999 there were about 500,000 Internet account holders in Malaysia, which would imply at least 1.5 million effective Internet users.

It is also worth noting that in Malaysia, PC-banking is more common among corporate customers compared to individual customers. This may be so since PC-banking has the advantage of reducing cost, increasing speed and improved flexibility of business transactions. To this extent, Leow(1999), asserts that on-line banking will emerge as a competitive tool and as a money saver rather than a revenue earner. However, if the banks are fast in adapting to the changing needs of consumers and are prepared to provide new innovative services in line with the demands of the electronic marketplace they may be able to transform this delivery channel into a revenue-generating source as well.

Automated Self Banking Centers

The automated self-banking center, is another multimedia banking delivery channel which incorporates an information counter, ATM's, telebanking and banking booths. These automated self-banking centers are usually situated in high pedestrian traffic areas such as shopping malls and office complexes. The Phileo Allied bank is the innovative leader in virtual multimedia bank kiosks in Malaysia with the PALVIRTUAL KIOSK, which consists of PALWORLD, PALPOINT, PALPHONE and PALTELLER touch screen terminals with video- conferencing capabilities.

PALWORLD allows users to log onto a world of online conveniences such as banking, electronic shopping, utility bill payment, share investment, flight bookings and hotel reservations and news updates including the Kuala Lumpur Stock Prices. PALPOINT is a new generation ATM facility that is connected to over 2800 ATM's via the GREAT and MEPS network nationwide. PALPHONE on the other hand, is the telebanking component of the virtual kiosk, which provides a variety of banking products and services via the Automated Voice System. Finally, the PALTELLER is a new generation terminal, which allows customers to conduct banking operations on-line via a touch-screen interface with video conferencing capabilities where the customer will be able to see and speak to a customer service officer located at the head office. The BSN Commercial Bank (M) Sdn Bhd., is another local Malaysian bank which, has also started to provide online banking services via the multimedia delivery channel. Through their electronic banking centers (EBC's)

Internet Banking

Internet banking would free both bankers and customers of the need for proprietary software to carry on with their online banking transactions. To this extent, it was pointed out by ABJ(1998), that five banks in Malaysia have invested millions of ringgit in Internet ready technology but these banks are still operating as dial up intranet facilities. In this context, an interview with a Malaysian banker, revealed that all home or PC-banking based on proprietary software which are browser based and conform to Internet requirements can also be used to provide Internet banking services. However, this delivery channel is still not available to Malaysian bank customers at present due to lack of adequate legal framework and security concerns. To this extent, EPI (1996), reported that one of the key factors which had contributed to the success of Security First Network Bank (SFNB), was the implementation of a comprehensive security infrastructure. The security infrastructure includes layers of security from the network to the browser, including sophisticated encryption that protects customers' from intrusion when they access the bank over the public network.

Furthermore, no matter how secure the bank, consumer confidence is paramount for the success of Internet banking. It is for this reason, that Internet banking is still not made available in Malaysia. The government wants the infrastructure both legal and physical to be in place before launching this additional delivery channel for financial products and services.

Other Recent Developments in Electronic Banking and E-Commerce in Malaysia

The role of banks in facilitating e-commerce is basically in providing the necessary electronic payment system, which is certainly a precondition for the success of e-commerce implementation. To this extent, the Malaysian Electronic Payment System (MEPS) Sdn Bhd, a major operator of ATM's in Malaysia and which is jointly owned by local commercial banks is in the process of investing more than RM350 million to set up an electronic emporium in an effort to expand its range of services to encompass e-commerce facilities for both businesses and individual customers, cyber-banking and a host of other electronic cash payment systems.

The immediate plan is to expand the availability of the MEPS cash electronic purse, which will become the country's first multi-purpose card. The card was first tested during the 1998 Commonwealth Games in Kuala Lumpur when the company issued 70,000 disposable e-purse cards. MEPS is now planning to study the acceptability of the card from a commercial aspect by launching the e-purse commercial pilot project in Bangsar Baru, Kulala Lumpur which would feature 21,000 reloadable and 75,000 disposable cards. The MEPS cash or e-purse which is targeted at retail merchants such as fast food outlets, restaurants, supermarkets and hypermarkets is the first of many applications which will eventually make up the Payment Multi-Purpose Card (MPC). An important benefit to the merchants is that the MEPS cash is cheaper than a credit card transaction. The long-term objective is to evolve the e-purse by the year 2002 into a widely usable multipurpose payment card that could serve as both a debit card and an ATM card.

The inter-bank GIRO system is intended to support the electronic presentation and payment of bills. MEPS's e-commerce initiative for banks in Malaysia was spearheaded with the launching of a pilot programme between banks and businesses in November 1998. However, only four banks and five retailers are currently participating in the e-commerce project but the number of merchants is expected to increase to 100 by the middle of year 2000. Finally MEPS is planning to expand its Secure Electronic Transaction (SET) Payment Gateway to offer Internet or Cyber Banking services on behalf of all banks in Malaysia.

This study on the evolution of electronic banking in Malaysia will not be complete without a proper analysis of the types of electronic delivery channels utilized by the local commercial banks. To this extent, the focus will be on the three main delivery channels namely ATM's, Telebanking and PC-banking. This information was basically gleaned from the commercial banks' web-sites and other brochures and pamplets.

The findings in Table 1 indicate that ATM's are undoubtedly the most popular electronic delivery channel for banking services in Malaysia. In regards to telebanking and PC-banking, though the Malaysian banks are clearly making the necessary efforts to provide these services, they have not penetrated the Malaysian market in a big way as yet.

Malaysian Banks' Use of the World Wide Web (WWW)

Diniz(1998), reported that a survey on bank web sites in the USA had indicated that American banks were using the Web to reach opportunities in three different categories: to market information, to deliver on-line banking products and services and to improve customer

Table 1: Electronic delivery channels utilized by local Malaysian banks

Commercial Banks	ATM's	Telebanking	PC-Banking
Malayan Banking Bhd.	a	a	a
Bank Bumiputra Malaysia Bhd.	a	a	r
Arab Malaysian Bank Bhd	a	a	r
Hong Leong Bank Bhd	a	r	a
The Pacific Bank Bhd	a	a	a
Ban Hin Lee Bank Bhd	a	r	a
Multi-Purpose Bank Bhd	a	a	a
Public Bank Bhd	a	a	r
Bank of Commerce Bhd	a	r	a
RHB Bank Bhd	a	a	r
BSN Commercial Bank Bhd	a	a	a
EON Bank Bhd	a	r	r
United Overseas Bank Bhd	a	a	r
Bank Islam Malaysia Bhd	a	r	r
Sime Bank Berhad	a	r	r
Sabah Bank Bhd	a	r	r
Oriental Bank Berhad	a	r	r
Southern Bank Berhad	a	r	a

a = does provide service
r = does not provide service

relationships. In this context, it is also worth noting that interactivity in the bank web sites is an important factor for the success of internet- banking services.

To this extent, an evaluation of the Malaysian commercial banks web sites or homepages revealed that most of these banks are basically using the Web at the brochure level merely for promotional purposes. However, some of the banks, which offer PC-banking services are probably exploring a wider aspect of the opportunities and possibilities of the WWW.

Another possibility the Malaysian banks should consider in relation to the use of the Web may be the provision of telebanking services via web phones, which may be cheaper than that via fixed line services.

Electronic Banking Issues

Although it is evident that the electronic revolution has commenced in Malaysia, widespread electronic banking may still be several years away. Star (Friday, September 3,1999, p12/Business) reported that a research survey conducted by PriceWaterhouse-Coopers on 40 to 50 senior executives of financial institutions based in Europe, North America and Asia Pacific had indicated that banks are not fully prepared for the 'paradigm quake' with the advent of information technology which is set to rock the banking industry. One of the main factors, which would be essential for the success of electronic banking, is setting up of the appropriate infrastructure. Once the infrastructure is in place then the commercial banks can start to push customers to accept the new delivery channels by ensuring that the necessary security measures are in place.

Next, even if the infrastructure were ready, the obvious question would be are the potential customers large enough to ensure a critical mass for the economic viability of providing the electronic banking services. If the critical mass criteria, is not fulfilled then the banks may not be able to profitably use these delivery channels. In the case of ATM's there is no doubt about the critical mass because almost every adult who has a bank account would have an ATM card and ATM's are available in every town, in every state in Malaysia. As far as telebanking is concerned, the existing four million fixed line telephone subscribers would imply that there is an adequate critical mass for telebanking services.

In the case of internet banking there are at present 500,000 Internet account holders in Malaysia and this is assumed to imply at least 1.5 million effective Internet users in the country which is about 7.5% of the Malaysian population. Can this be considered as a sufficient critical mass for harnessing the new delivery channel? If the critical mass has not been reached, should the banks wait until it has before offering their products and services through the new delivery channels or should the banks introduce the new delivery channels and persuade or push the customers to use them? These are questions that need to be answered in implementing electronic banking not only in Malaysia but also in any part of the world.

According to an article, gleaned from the web, it was reported that a Cybercitizen Finance Study by Cyber Dialogue had indicated that as at end of 1998 though the number of consumers banking online had grown to 6.3 million, 3.1 million US adults had also discontinued their use of online banking. More than 50% of those who had discontinued, found the service too complicated or were dissatisfied with the level of customer service. This indicates that for successful implementation of electronic banking in Malaysia, providing banking products and services through the new delivery channels alone is not sufficient. The banks should in fact ensure that consumption of the services through these new delivery channels are simple, easy and of sufficiently high quality to ensure customer satisfaction in order to maintain their online customers. This is going to be all the more difficult because of the lack of personal touch in electronic banking.

Another problem associated with electronic banking is electricity power cuts, which can be disruptive. What will happen to transactions, which are being processed when a power cut occurs. Thus, one infrastructure requirement that is essential from the outset is stable power supply, which is not a basic feature in most, less developed and developing countries including Malaysia.

In addition, to power stability is the issue of connectivity. In some countries the local internet service providers may not be very efficient in terms of connectivity. An article from Electronic Payments International (August 1996), reported that when America On Line (AOL) shut down for 19 hours on August 7 th 1996, business users were unable to surf the net, send electronic mail and make online payments. To this extent, many information publishers complained of losing thousands of dollars in access fee and advertising revenues.

Thus, for the successful implementation of electronic banking in Malaysia or in any other part of the world, adequate legal and physical infrastructure are major prerequisites. Then, the customers must also be made to feel confident about the privacy and security issues associated with electronic banking. Finally, quality service would certainly be a very important determinant for the success of electronic banking.

Survey Findings on Consumers' Perception of Electronic Banking in Malaysia

The findings reported in this section are based on data collected for a pilot study on consumers' perception of electronic banking in Malaysia. The respondents for the pilot study primarily consisted of employees of the Multimedia University, which is located in Cyberjaya, at the heart of Malaysia's Multimedia Super Corridor (MSC). The pilot study only involved thirty respondents but the final survey will be based on a sample of 300 bank customers. The data for the pilot study was collected by administering, the personal questionnaire method. Though the questionnaire covered various issues only some of the relevant findings are reported here.

Table 2: Frequency of visits to bank branche

Number of Visits to Bank Branch Every Month	Frequency	Percentage(%)
Never	3	10
Once	14	46.7
Twice	6	20
Three or more	7	23.3
Total	30	100

Table 3: Frequency of ATM usage

Number of Usage per Month	Frequency	Percentage(%)
Once	1	3.3
Twice	6	20
Thrice	4	13.3
Four or more	19	63.3
Total	30	100

Table 4: Importance of ATM Usage

Importance of Human Tellers	Frequency	Percentage(%)
Very Important	6	20
Important	21	70
Not Important	3	10
Total	30	100

One of the implications of electronic banking is that it should reduce the need to visit bank branches. In fact, the electronic banking delivery channels are often considered as potential substitutes for brick and mortar bank branches. However, contrary to expectations, the survey results indicated that 90% of the respondents visit their bank branch at least once every month.

This would indicate that consumers of banking services in Malaysia still find it useful to visit their bank branches regularly every month to perform some banking transactions such as mortgage and loan repayments for which the automated payment systems are not very common. However, these customers also indicated that they frequently used the ATM's for other transactions such as cash withdrawals and funds transfer. To this extent, 63.3% of the bank customers indicated four or more visits to the ATM's every month.

Furthermore, 90% of the sample respondents also indicated that they find human tellers important. This would indicate that bank customers in Malaysia do quite highly value the importance of the personal touch in banking services.

In relation to the usage of the three main electronic delivery channels for banking services which have already been in existence in Malaysia, namely Automated Teller Machines, Telebanking and PC-banking, the survey results indicated that ATM's are the most widely used electronic banking facility.

However, only 20% of the respondents were using telebanking services and only one respondent accounting for 3.3% of the respondents was familiar with PC-banking and personal finance management software.

Though internet-banking is not available in Malaysia at the moment about 36.7% of the respondents indicated that they would

Table 5: Familiarity and usage of electronic banking channels

Delivery Channel	ATM's		Telebanking		PC-Banking	
	Frequency	%	Frequency	%	Frequency	%
Currently In Use	30	100	6	20	1	3.3
Not In Use	0	0	24	80	29	96.7
Total	30	100	30	100	30	100

Table 6: Internet availability at home

Internet Access	Number of Respondents	Percentage(%)
Available	19	63.3
Not Available	11	36.7
Total	30	100

be very interested in using this service once it is made available and only 6.7% of the customers indicated that they would not be interested at all in using this service.

Furthermore, 63.3% of the customers surveyed indicated that they had Internet access at home. This figure may represent an exaggeration of the national situation since the survey respondents are basically university staff who are quite highly computer literate and hence may not be representative of the whole nation. However, the attitudes of this group might be indicative of future trends in the Malaysian population. Nevertheless, the findings also indicate positive potentials for the introduction of Internet banking in Malaysia.

Conclusion

Advances in information technology and telecommunications have certainly introduced new delivery channels for Malaysian commercial banks' products and services. These new delivery channels include automated teller machines (ATM's), telebanking via the telecommunication channel, and PC-banking based on proprietary software. Among these, the ATM's are the most widely accepted and highly utilized delivery channel. Telebanking has not been very successful in Malaysia, but PC-banking has been successful to some extent among corporate customers. However, the indications are that the PC-based channels of banking have not realized its full potential in Malaysia.

Internet banking is still not available in Malaysia. It may seem ironical that a country, which has undertaken great initiatives in supporting developments on the Internet, still disallows its banks from providing Internet banking services. The governments main concern is the security of these electronic transactions.

The general survey findings are that, most Malaysian banking customers still patronize the bank branches and find interaction with human tellers as important. However, over 60% of the respondents have Internet access at home and these represents a positive indication for PC-based banking and Internet banking in the future.

The data for this study was primarily obtained from a pilot study on consumer perceptions of online banking products and services. Thus, the sample suffers from two limitations. First, the sample size of only thirty respondents and second the sample respondents who were mainly employees of the Multimedia University. Despite these shortcomings the results compared quite well with that of a larger sample study carried out by BSN Commercial Bank. Thus, the findings of this study are not seriously flawed by this limitation.

Biography: Balachandher Krishnan Guru
Faculty of Management, Multimedia University, Cyberjaya, 63100, Selangor, Malaysia
e-mail: bala@mmu.edu.my

Santha Vaithilingam Faculty of Management, Multimedia University, Cyberjaya, 63100, Selangor, Malaysia
e-mail: santha@mmu.edu.my

Norhazlin Ismail
Faculty of Management, Multimedia University, Cyberjaya, 63100,Selangor, Malaysia
e-mail: norhazlin.ismail@mmu.edu.my

Rajendra Prasad
BSN Information Technologies & Services Sdn. Bhd, Kelana Business Center, Malaysia
e-mail: prasad@bsnit.com.my

Chapter 3: Secure Banking

Security White Paper, A Closer Look

Title: *Security White Paper, A Closer Look*
Author: *Security First Network Bank*
Abstract: *How do banks ensure the safety of their transactions? The Security First Network Bank uses all the possible tools to this effect: cryptography, firewalls and routers, trusted operating system and internal control.*

Cryptography

Security First Network Bank utilizes several layers of technology to ensure the confidentiality of its transactions across the Internet. Security begins with your browser. Netscape's SSL protocol (Secure Sockets Layer) is used to provide privacy for the data flowing between the browser and the bank server.

SSL is an open protocol for securing data communication across computer networks, and it provides a secure channel for data transmission through its encryption capabilities. It allows for the transfer of digitally signed certificates for authentication procedures, and provides message integrity, ensuring that the data can't be altered en route.

When a customer account is created, the bank assigns a password which is sent to the customer along with an account verification letter. In addition to password protection, Security First Network Bank also provides server authentication using the latest in public key cryptography.

Public/private key pairs are used specifically for authentication. The public key can be distributed, using a certificate that verifies the identity of the owner. The private key is kept secret. A message encrypted with a public key can only be read after decryption with the private key.

To start a transaction, the customer uses his or her browser to send a secure message via SSL to the bank. The bank responds by sending a certificate which contains the bank's public key. The browser authenticates the certificate, then generates a session key which is used to encrypt data traveling between the customer's browser and the bank server.

The session key is encrypted using the bank's public key, and sent back to the bank. The bank decrypts this message using its private key, and then uses the session key for the remainder of the communication.

By exchanging messages using the public/private key pair, the customer can be assured they are actually communicating with the bank, and not a third party trying to intercept the transaction. When a session is encrypted, the key icon at the lower left corner of the browser's screen becomes solid, and a blue line appears at the top of the screen. If the key icon appears broken, encryption is not in use and the current session is not secure.

Firewalls and Routers

Security First Network Bank has gone to great lengths to ensure that your money and personal data are protected against any type of intruder or attack.

The bank is protected by a system of filtering routers and firewalls, which form a barrier between the outside Internet and the internal bank network. The filtering router verifies the source and destination of each network packet, and determines whether or not to let the packet through.

Figure 1

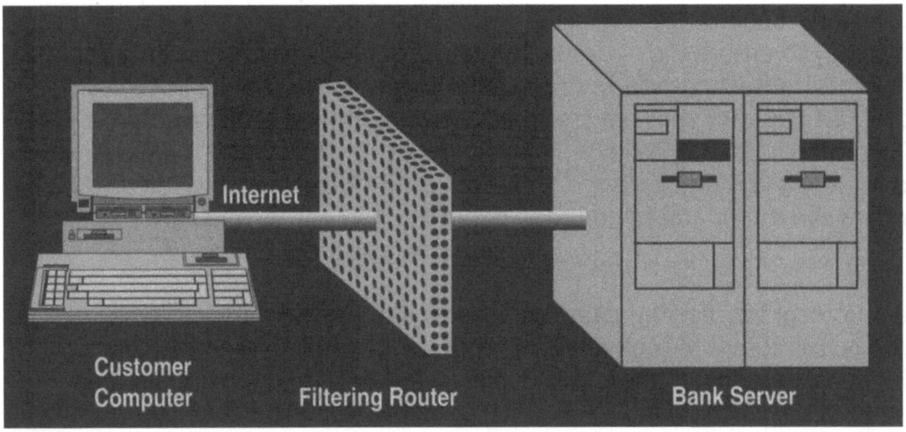

Figure 2

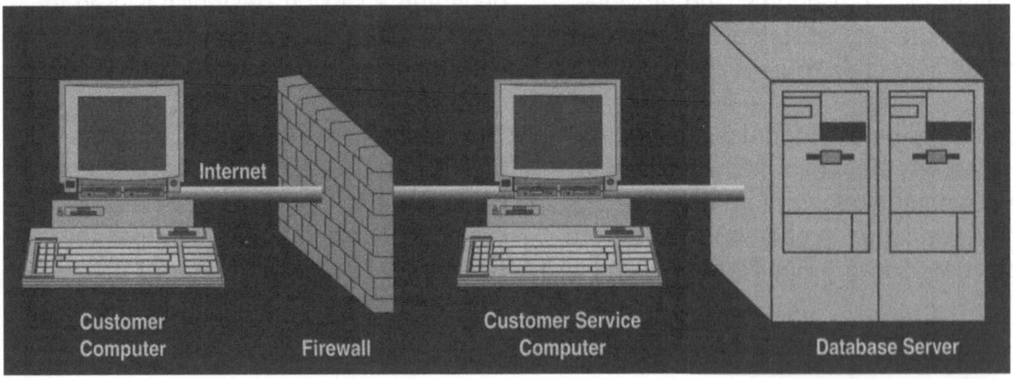

Access is denied if the packet is not directed at a specific, available service.

The firewall is used to shield the bank's customer service network from the Internet. All incoming IP traffic is actually addressed to the firewall, which is designed to allow only e-mail into the customer service environment. Traffic through the firewall is subjected to a special proxy process which operates in much the same way as a filtering router, verifying the source and destination of each information packet.

The proxy then changes the IP address of the packet to deliver it to the appropriate site within the customer service network. In this way, all inside addresses are protected from outside access, and the structure of the bank's internal networks is invisible to outside observers.

The Trusted Operating System

While there are important security issues associated with transit across the Internet, the greatest risk to your financial information occurs within the bank itself. Security First Network Bank addresses this issue using Security First Technologies Virtual Vault platform. An important part of this architecture is the Trusted Operating System, the dominant security platform in government computing.
Security First Network Bank's use of this trusted operating system, called CMW+, represents the first commercial implementation of this highly successful platform, used for years by the Department of Defense and other high-security government agencies.

The trusted operating system acts as a "virtual vault," protecting customer information and funds inside the bank. It uses multilevel technology and contains privilege and authorization mechanisms to control access to functions and commands. It also contains an audit mechanism which records logins and logouts, use of privilege, access violations and unsuccessful network connections. This allows quick identification of any suspicious activity.

Figure 3

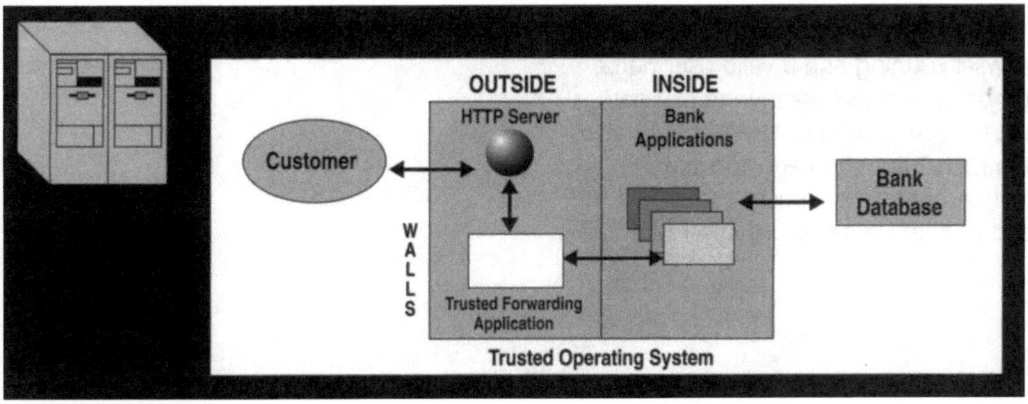

Internal Controls and Customer Responsibility

Strict internal procedures are in place within Security First Network Bank Network Bank, controlling every aspect of bank administration from training employees to confirming customer transactions to preventing service interruptions.

New advances in security technology are happening daily, and Security First Network Bank is constantly evaluating its security architecture to ensure that it provides the highest level of privacy and safety for bank customers. For more information about the bank's security architecture and plans for future releases, see our white paper titled, A Closer Look: Security First Network Bank's Security Architecture.

Customers have their own set of responsibilities in providing security for their Internet bank account.Passwords must be kept secret. Users should make sure that no one is physically watching as passwords are entered. It is important to remember to exit the browser when leaving the computer.

If the PC is left unattended with the browser running and a valid user name and password cached,anyone can gain access to the account. Users should also take precautions to keep computers cleanand free from viruses that could be used to capture password keystrokes.

Server Gated Cryptography: The Solution to Secure Worldwide Banking

Title: *Server Gated Cryptography*
Author: *Microsoft*
Abstract: *Server Gated Cryptography (SGC) empowers financial institutions with Microsoft® Windows NT® based Internet servers to protect online banking sessions using strong 128-bit encryption. SGC is included as part of Microsoft's standard, exportable versions of its Web browser and Web server software, and is fully compliant with all U.S. government export regulations. Additional information is available on the Microsoft SGC Web page.*

Overview

Until now, strong encryption products have been unavailable to the worldwide banking industry. U.S. export laws regarding cryptographic software have prevented banks and other financial institutions from providing security to their World Wide Web services using the high-security products that are available in North America. However, SGC offers strong encryption that is available worldwide.

SGC is an extension to the well-known Secure Sockets Layer (SSL) protocol that is used to secure Web sessions. Unlike SSL, which must be produced in both an exportable version that uses a 40-bit cryptographic key and a North American version that uses a 128-bit key, a single version of SGC can be produced, one that uses a 128-bit key worldwide. This can be done because of a recent U.S. Department of Commerce decision to license the export of strong cryptographic products in cases where it is possible to restrict them to financial institutions only.

SGC is included in all of Microsoft's current Web client and server software, and previous versions of these products can be upgraded by simply downloading a file from Microsoft's SGC Web page. Using SGC, financial institutions can provide secure Web services for their customers, without the need to change Web browser or server software. Microsoft's SGC product is fully compatible with Netscape's SGC product.

How SGC Works

SGC is a technology that enables Web client and server software to dynamically determine the encryption level of an SSL session. It is implemented as an extension to the Secure Socket Layer (SSL) protocol and its successor, the Transport Layer Security (TLS) protocol. To use SGC, both the client and server must be running software that implements SGC, and the bank must have an SGC digital certificate from a certificate authority (CA).

When a customer connects to an SGC-secured Web site, the browser software and the Web server software initiate a normal SSL connection. During the "handshake" portion of the session, the client checks the server's digital certificate and looks for special data in the certificate that indicates that the server can participate in an SGC session. If the special data is not in the certificate, the client negotiates a normal SSL session, using a 40-bit key. However, if the special data is in the certificate, the client resets the handshake and negotiates a session using one of the following cryptoalgorithms and a corresponding key:

- 128-bit RC4
- 128-bit RC2
- 56-bit DES
- 3Key 3DES (pending final approval by the U.S. Department of Commerce)

SGC Digital Certificates

The bank must have a valid SGC digital certificate installed in its Web server software in order to use SGC. To obtain a certificate, a bank must prove that it is a bona fide financial institution. This is usually done by providing a Dun & Bradstreet DUNS number or an American Banking Association number, but also can be done by submitting documents from a state, provincial, or national government verifying that the institution has been chartered as a bank. This level of proof is necessary in order to ensure that the provision under which the U.S. Department of Commerce granted license to export SGC – namely, that it be used only by financial institutions – is being met. A full listing of the CAs that provide digital certificates for use with SGC is available on the Microsoft SGC Web page. Once issued, the certificate is installed in exactly the same way as an SSL certificate.

Availability

SGC is included in the Windows 95 and Windows NT versions of Microsoft Internet Explorer 4.0 and Microsoft Money 98, as well as Microsoft Internet Information Server (IIS) 4.0, which works only with Windows NT Server. Users who have Internet Explorer 3.02 or IIS 3.0 can download a free upgrade from Microsoft's SGC Web page. In addition, Microsoft will include SGC in future releases of the 16-bit Windows 3.1-based version of Internet Explorer and in the Macintosh and Unix versions.

Interoperability

Microsoft has worked to ensure that its implementation is fully interoperable with Netscape browsers and servers. This means that banks using IIS can communicate with customers who use both Netscape and Microsoft Web browsers, and customers who use Microsoft Internet Explorer can communicate with banks that use Netscape Web servers. Strong cryptography is employed in each case.

Eligibility Requirements

SGC is generally available to qualified banking and financial institutions worldwide. Exceptions include banks and financial institutions that are located in countries subject to a U.S. embargo (a partial list includes Cuba, Iran, Iraq, Libya, North Korea, Syria and Sudan) and several additional restricted destinations (a partial list includes Russia and the People's Republic of China).

Web Site Authentication: Community Banks, SiteCertain and On-line Banking

Title: *Website Authentication: Community Banks, Site Certain and Online Banking*
Author: *ABAecom*
Abstract: *This report provides answer the following questions:*
Why should I concern myself with on-line banking? How can I provide greater assurances to my customers that my Web site is secure? What is SiteCertain and how much does it cost?
What are Digital Certificates? What are some other activities that ABAecom has underway?

Executive Summary

ABAecom was established to facilitate electronic commerce by creating a secure environment for financial institutions and their customers. Customers won't use the Internet for financial transactions unless they can verify that they are dealing with a legitimate financial institution and that the information they are sending will be secure. ABAecom believes the use of public key cryptography and digital certificates are crucial to the growth of on-line financial and commercial transactions.

ABAecom will act as a root certification authority, assuming a trusted third party position, and will issue Digital Certificates to legitimate financial institutions. ABAecom will also offer services to help enable banks to issue Digital Certificates. Our goal is to provide privacy, authentication, data integrity and non-repudiation for Internet transactions and communications through a wide range of services and education. This will include the involvement of banks as trusted entities using digital certificate technology to vouch for the electronic identity of businesses and individuals.

This report provides answer the following questions:

- Why should I concern myself with on-line banking?
- How can I provide greater assurances to my customers that my Web site is secure?
- What is SiteCertain and how much does it cost?
- What are Digital Certificates?
- What are some other activities that ABAecom has underway?

The Trends

Market statistics regarding electronic commerce and on-line banking are becoming harder to ignore. While still representing a small percentage of the total economy, the Internet is nonetheless affecting every industry and encompassing business-to-business commerce as well as retail. The global economy is migrating to the Internet at a dizzying rate.

Recent articles have proclaimed the following:

- Forrester Research: Internet commerce in the United States will reach $127 billion in 1999 and almost $1.5 trillion in 2003.
- "The Internet is shaping up as the battleground of the future for financial institutions competing for small business customers."

  Jennifer Weitzman, American Banker On-line, 11/6/98.
- "While bankers may be preoccupied with the Y2K bug, their best customers are looking for sophisticated electronic access to their financial information. For this reason, community banks need to get into the Internet banking services game, and 2000 is a clear goal post for banks striving to do it."

  David Peterson, Independent Banker, 2/99.
- "The mainstreaming of e-commerce will irrevocably alter the financial services landscape."

  Ernst & Young LLP, "E-commerce & Connecting to the Customer," 1998

The "pitch" goes something like this: With electronic commerce, banks can increase revenues, decrease cost of operations, attract new customers, embed existing customer relationships, and develop new offerings.

Some banks may disagree. Others may agree, though believe such a reality is far off the horizon. Given, however, the tremendous growth rate of commerce over the Internet--everything from on-line banking, to brokerage accounts, to government permits, to book sales, to business-to-business trading--it is unlikely that e-commerce will wait for the banks to stake their claims in "cyberspace." New companies, new alliances, new business models, all continue to creep into areas traditionally dominated by banks. The Internet rages on, commoditizing services, products, and even whole industries. ABAecom believes that service, trust, brand, reputation, capital, knowledge of customers, their needs, habits, desires, will not be commodities and are qualities that can make banks essential players in the electronic world.

According to Faulkner & Gray's 1999 Guide to Retail Banking on the Internet, of the 700 smaller banks ($1 billion assets or less) that actually have Web sites, only about 20% of those have sites that offer transactional capabilities, compared to about 45% of banks with more than $6 billion. Of course, many consider the Internet the "great equalizer," and those community banks moving quickly into the world of on-line banking will be able to offer Internet services like much larger banks, while still maintaining their existing strengths: superior customer service and knowledge of their customers. On the Internet, where any site, any business, any person, merchant, entity, is only a click away, fundamental trust upon a community-based institution, that knows and serves its customers well is a prime asset. This particular relationship can be instrumental to eliminating customers' uncertainty for conducting their business activities over the Internet.

Recent news seems to hint that a community bank, capitalizing on its strengths in the physical world, and transporting them to the virtual world, offers a potent combination. An article in the February 1999 issue of Future Banker profiled Salem Five Cents Savings Bank ($1 billion assets) CEO William Mitchelson. "We have an experience level that many big banks lack because we entered the field so early," he said. Continuing, he noted that "all banks should be on-line. That's because customers dictate how they will bank, not banks."

According to Forrester Research, "in another 18 months to two years, providing Internet access will be a decisive factor for small businesses to move [a banking relationship]." Despite the enormous marketing efforts that very large banks may be willing to undertake, it remains to be seen whether such an effort will sway business customers who want the convenience and service from a community bank branch. Nonetheless, as on-line commerce becomes routine—and expected—it is highly likely that businesses will no longer rely on a bank that does not offer top-notch on-line financial services.

On-line banking allows banks to retain customers, and to gain new ones. It is becoming a channel that businesses and consumers are demanding—expecting—and banks are attempting to meet this challenge. According to IDG News Service, sales of Internet banking applications reached $93 million in 1998 and "will jump some 250 percent to more than $326 million in 1999." The February 1, 1999 article noted that "more than 1,200 US banks and credit unions signed with on-line-banking application vendors and providers to build transactional sites on the World Wide Web in 1998, while (in 1999) 7,200 more will acquire on-line banking applications."

Beyond this, the sheer rate of growth in the Internet may be too enticing to ignore. Telebanc Financial Corp., a $2.3 billion-asset bank, which exists through the Internet and over the telephone, saw its fourth-quarter revenues rise 120% from the previous year. According to the American Banker, "exhibiting attributes more akin to an Internet company than a community bank," Telebanc saw its share price rise 10% the day it announced its earnings, in early February. The article continued: "Telebanc is growing in an industry that is stagnant. Internet banking will probably be the hottest sector on Wall Street in 1999 and 2000." The company spent $4.6 million on marketing in 1998. Telebanc has also reached a co-marketing arrangement with Internet portal, Yahoo.

Despite this, results of the ABA Banking Journal "'99 Community Bank Competitiveness Survey" revealed some surprising conclusions. For example, while the number of banks on the Web has doubled since 1997, it is biased towards the larger banks. Moreover, the survey found the typical community bank Web site was primarily used for non-banking services such as community information, e-mail to bank staff and financial calculations. Eighty percent of community banks with Web sites were receiving fewer than 500 hits per month.

Banks are historic trusted parties. On-line banking and electronic commerce will require an extension of that trust if they are to truly flourish. Banks provide security, accountability, predictability, certainty, performance, relationships, trust and value. As more individuals and businesses go on-line, they will demand that their bank goes on-line. That said, many customers expect the same security in the on-line world as in the physical world. Indeed, many bank Web sites go under-utilized as a large percentage of customers remain hesitant to trust any aspect of their banking relationship to the Internet. ABAecom, through its SiteCertain product, offers a solution to these quandaries.

SiteCertain

ABAecom's first product, SiteCertain, is a seal that will appear on a bank's Web page; "clicking" on the icon will provide simple instructions that enables customers to confirm that they have reached a legitimate financial institution—and not simply a site that has been "spoofed" or merely claiming to be "XYZ bank." SiteCertain is powered by public key encryption and Digital Certificates—more on that below.

Not just anyone can use the seal on their site. Prior to issuing the SiteCertain seal, ABAecom undertakes an in-depth identification and authentication process designed to ensure that the bank and its Web site are legitimate. Customers will appreciate the extra step that the bank has taken to protect them. And because customers will feel more comfortable, they will use the bank's Web site more frequently, boosting transaction volume, building trust and strengthening their relationships with the institution.

Here's how it works. The bank places the seal on one or more "pages" of their Web site. The customer sees the 'spinning' seal and is requested to "click" to verify that this is really a legitimate financial institution. The user is provided simple instructions on how to affirm, through the use of digital certificates, that they are at the site of a real bank, not an impostor site. The instructions also provide links to additional information about the bank, including a link to the FDIC database.

ABAecom's service provider, Digital Signature Trust (DST), provides the digital certificates behind the SiteCertain seal and maintains the public certificate directory. As a subsidiary of a national bank, Zions Bank, DST understands the security requirements and business needs of banks and bank customers.

How much does SiteCertain cost?
SiteCertain is free to ABA members. There are several reasons for this:

1. ABAecom is committed to maximizing the value of ABA membership.
2. ABAecom is committed to on-line banking and Internet commerce.
3. SiteCertain is designed to position financial institutions as the premier

providers of trust and security on the Internet.

4. We expect that by providing the service for free, and distributing it to legitimate financial institutions, the value of SiteCertain, to both banks and their customers, will be significantly enhanced.

The more often users see the seal, and understand the value of security and trust that banks can provide for Internet commerce, financial institutions will be viewed as the keystone to a new world of on-line transactions. Indeed, as will be discussed below, ABAecom is aggressively seeking to expand the opportunities for financial institutions in cyberspace, particularly through the use of digital certificates widely distributed to the general public.

Digital Certificates

Okay, so what is a digital certificate? A digital certificate is basically a very long number that clearly identifies you on the Internet, the way a driver's license and a handwritten signature, for example, would identify you in the physical world. It is more secure and easier to use than passwords. With respect to SiteCertain, a digital certificate allows for a secure session between the customer and the bank via its Web site.

Digital certificates, however, can provide far more. To understand digital certificates, though, first we must review something called public key cryptography. Public key cryptography is based upon the concept of a "key pair," a public key and a private key. Each key—a large prime number—can encrypt information that only the other key can decrypt. Only its owner knows the private key. The public key is widely distributed—that is, made available to all who wish to transact business with a party—via an entity called a Certificate Authority (CA). The public key resides within a computer file that is digitally signed by a Certificate Authority—this file is called a digital certificate. The Certificate Authority, then, vouches for the validity of a certificate making it safe to transact business with the person, company or Web site "at the other end of the line." In the case of SiteCertain, the CA is ABAecom, and the certificate holder is the bank. The digital certificate, delivered with SiteCertain, is held by the bank (on its Web server) and can be viewed by the public through their browser (Netscape and Internet Explorer), providing customers with the assurance that they can conduct their transactions securely. A certificate can be issued to a person, a business, and to an Internet server, and is typically issued for a stated period, oftentimes one year, as is the case with SiteCertain.

Technology companies, credit reporting firms, telecommunications firms, and the federal government are already positioning themselves as Certificate Authorities. It is assumed that most participants in trusted electronic commerce, including on-line banking, secure e-mail, secure documents, and other confidential Internet transactions, will only respect the validity of a public key if it is issued by a respected Certificate Authority. Banks, naturally, can play such a role.

ABAecom Activities

As noted above, ABAecom was created to facilitate electronic commerce by creating a secure environment for financial institutions and their customers. The following are some of the activities that ABAecom is engaged in:

Portable Certificate
ABAecom's value proposition has always been tied to the assumption that banks should be the purveyors of trust in an on-line world. Our goal is to build a standards based, interoperable system that will support the use of bank issued certificates across a variety of platforms and applications. We view this notion of "portability" (the ability of a bank issued certificate to be used in sectors as varied as government, healthcare and finance) as a key to our future successes.

On the Internet, a "portable cert" could allow an individual to transact business with his or her bank, broker, accountant, the DMV, IRS, a retail store, or some newly emerging entity—with bank certs providing the fundamental building blocks: trust, identification, authentication, and non-repudiation. For high-value and/or highly confidential on-line transactions, a secure form of identification will be required and banks are the logical choice to provide such a service. The Portable Cert envisions a world of electronic commerce in the new millennium. A world where you can complete high-value financial transactions in total privacy and with assurance that your business will be conducted exactly as directed, and instantly—think of it as a combination driver's license and passport. To obtain such a certificate, an individual will apply for one at their local bank.

Joint marketing materials
Currently, ABAecom provides materials such as a SiteCertain brochure, Web site pages, and a customer service number to assist banks (and their on-line customers) better understand Internet security and SiteCertain. To help support your on-line banking efforts and the added security of SiteCertain, ABAecom is developing additional Web-based and hard copy collateral that can complement bank efforts to promote secure on-line banking. ABAecom and DST are also planning a series of banking industry seminars to highlight the opportunities provided by digital certificates.

Policy Advisory Committee
One of the objectives of ABAecom is to determine and make recommendations on standards and policies in the general area of electronic commerce. We have established a Policy Advisory Committee composed of bankers as well as representatives from trade associations and regulatory agencies. The ABAecom Policy Advisory Committee, acting on behalf of the financial services industry, shall propose and craft policies relating to public key infrastructure technologies and electronic commerce applications.

New Web site
Premiering mid-February, 1999, the new ABAecom Web site (www.abaecom.com) will serve as a "portal" to the Internet banking community and their customers, offering insight and information on on-line banking, digital certificates and their role in

secure electronic commerce. Complete registration and enrollment information for SiteCertain is available through this site.

Entrust Worldwide
ABAecom is a founding participant in Entrust Worldwide. This is an effort by a major technology company, Entrust Technologies, to establish a global network of Certificate Authorities. These CA's will agree to establish a set of rules to enable the exchange of e-mail between members of the network. The goal is to create a system of trusted e-mail that can be relied upon the same way as paper-based business correspondence is today.

Public Technology, Inc. (PTI)
PTI is the technology subsidiary of the National League of Cities, the National Association of Counties and the International City/County Management Association. ABAecom and PTI recently issued a press release that outlined our joint effort to educate our respective members and the public on the benefits of secure electronic commerce and PKI. This relationship will involve joint communication efforts as well as pilots between local governments and financial institutions using digital certificates.

ABAecom was established to facilitate electronic commerce by creating a secure environment for financial institutions and their customers. Customers won't use the Internet for financial transactions unless they can verify that they are dealing with a legitimate financial institution and that the information they are sending will be secure. ABAecom believes the use of public key cryptography and digital certificates are crucial to the growth of on-line financial and commercial transactions. ABAecom will act as a root certification authority, assuming a trusted third party position, and will issue Digital Certificates to legitimate financial institutions. ABAecom will also offer services to help enable banks to issue Digital Certificates. Our goal is to provide privacy, authentication, data integrity and non-repudiation for Internet transactions and communications through a wide range of services and education. This will include the involvement of banks as trusted entities using digital certificate technology to vouch for the electronic identity of businesses and individuals.

For more information please visit our Web site at www.abaecom.com, or send e-mail to bhall@aba.com. We can be reached via phone at 202.663.7584.

Risk and Prevention in Web-Based Financial Applications (a CustomerLink Security White Paper)

Title: *Risk and Prevention in Web-Based Financial Applications (a CustomerLink Security White Paper)*

Author: *Goldleaf Technologies, an Equifax Company*

Abstract: *The Internet is designed as an open system based on the premise of free access and communication between participating computer systems. While nothing can guarantee complete safety and invulnerability, every effort must be made to provide a safe, secure, and protected platform for financial activity.*

Overview

The incredible growth of the Internet is fostering a rapid movement of financial institutions into the electronic arena. In a recent survey, the majority of financial institutions said they were in the process of implementing, or planning to implement some form of Internet presence by the beginning of the next century. The virtual stampede into this new marketplace is being slowed only by a widespread concern for safety and security as confidential information and sensitive payment instructions are launched into the uncharted sea of the World Wide Web.

Conversations about the opportunities and pitfalls of "Internet Banking" are to be found in almost every trade magazine and industry forum. This document will attempt to condense the wisdom of those groups for whom the study of the Internet as a vehicle for commercial activity is a full-time occupation. The goal is to educate financial institutions considering the movement toward electronic services of the risks that do exist, and the measures that can be taken to limit those risks. Where applicable, the appropriate link to more information has been provided.

Security Risks

Each financial institution that establishes an Internet presence must take proactive steps to ensure that the systems they own or participate in contain adequate security measures to limit the possibility of unintended distribution of confidential information and the potential for fraud-related losses. The Internet is designed as an open system based on the premise of free access and communication between participating computer systems. While nothing can guarantee complete safety and invulnerability, every effort must be made to provide a safe, secure, and protected platform for financial activity.

According to the Federal Deposit Insurance Corporationís white paper entitled "Security Risks Associated with the

Internet" (FDIC Division of Supervision, December 1977), there are five primary areas of risk associated with the provision of financial institution data and the initiation of financial transactions on the Internet.

From the FDIC's "Security Risks Associated with the Internet":

Data Privacy and Confidentiality
Unless otherwise protected, all data transfers, including electronic mail, travel openly over the Internet and can be monitored or read by others. Given the volume of transmissions and the numerous paths available for data travel, it is unlikely that a particular transmission would be monitored at random. However, programs, such as "sniffer" programs, can be set up at opportune locations on a network, like Web servers (i.e., computers that provide services to other computers on the Internet), to simply look for and collect certain types of data. Data collected from such programs can include account numbers (e.g., credit cards, deposits, loans) or passwords.

Due to the design of the Internet, data privacy and confidentiality issues extend beyond data transfer and include any connected data storage systems, including network drives. Any data stored on a Web server may be susceptible to compromise if proper security precautions are not taken.

Data Integrity
Potentially, the open architecture of the Internet can allow those with specific knowledge and tools to alter or modify data during a transmission. Data integrity could also be compromised within the data storage system itself, both intentionally and unintentionally, if proper access controls are not maintained. Steps must be taken to ensure that all data is maintained in its original or intended form.

Authentication
Essential in electronic commerce is the need to verify that a particular communication, transaction, or access request is legitimate. To illustrate, computer systems on the Internet are identified by an Internet protocol (IP) address, similar to a telephone that is identified by a phone number. Through a variety of techniques, generally known as "IP spoofing" (i.e., impersonating), one computer can actually claim to be another. Likewise, user identity can be misrepresented as well. In fact, it is relatively simple to send an e-mail message that appears to have come from someone else, or even send it anonymously. Therefore, authentication controls are necessary to establish the identities of all parties to a communication.

Non-Repudiation
Non-repudiation involves creating proof of the origin or delivery of data to protect the sender against false denial by the recipient that the data has been received or to protect the recipient against false denial by the sender that the data has been sent. To ensure that a transaction is enforceable, steps must be taken to prohibit parties from disputing the validity of, or refusing to acknowledge, legitimate communications or transactions.

Access Control/System Design
Establishing a link between a bank's internal network and the Internet can create a number of additional access points into the internal operating system. Furthermore, because the Internet is global, unauthorized access attempts might be initiated from anywhere in the world. These factors present a heightened risk to systems and data, necessitating strong security measures to control access. Because the security of any network is only as strong as its weakest link, the functionality of all related systems must be protected from attack and unauthorized access. Specific risks include the destruction, altering, or theft of data or funds; compromised data confidentiality; denial of service (system failures); a damaged public image; and any resulting legal implications. Perpetrators may include hackers, unscrupulous vendors, former or disgruntled employees, or even agents of espionage.

Stephen Cobb, Director of Special Projects for the National Computer Security Association, summarized these same components in the following quote from the paper "Security Issues in Internet Commerce":

"The challenge is to transmit and receive information over the Internet while insuring that it is inaccessible to anyone but sender and receiver (privacy), it has not been changed during transmission (integrity), the receiver can be sure it came from the sender (authenticity), the sender can be sure the receiver is genuine (non-fabrication), and the sender cannot deny he or she sent it (non-repudiation)."

Types of Attacks

Considering these areas of potential risk, Web site managers must wage the battle to secure their systems along a very wide front. Successful security requires more than just knowledge of the danger areas; it also depends on the site manager's ability to identify the avenues through which security threats may access internal systems. According to the International Computer Security Association's "Web Site Security: Adoption, Risk & Prevention" (ICSA Release 1.3 -- November 1997, http://www.goldaero.com/DemoHoliday/www.icsa.net) Web site managers must contend with an unsettling array of potential problems:

Web-Based Attackers
Web-based attackers, more commonly called "hackers," pose the most publicized security challenge. These individuals gain access to systems and the information they contain by exploiting flaws in the configuration of the Web server, the server's operating system, or the actual components of the Web pages.

Internal Intrusion
Internal intrusion, or the breaching of security systems by someone within the organization who might potentially have authorized access to hardware and software components, is another area of concern. Physical security and password security are paramount in controlling illegal activity within the site manager's own organization.

Malicious Code
The receipt or distribution of malicious code elements can also cause security risks and, more commonly, malicious harm to Web-based computers and programs. Many Web sites include receipt and distribution of files and programs among the features that are offered to end users.

Denial of Service
Denial of service and availability must also be a primary concern for site managers. As if protection from interlopers and subversive elements was not enough of a concern, system downtime resulting from power failure, wide-area communications problems, and natural disasters can also result in customer dissatisfaction and ultimate harm to the site manager's company or group.

Negligence
Security-related negligence in the handling of sensitive data or the configuration of security systems can also be devastating. The best security systems incorrectly administered or improperly managed present increased risk potential. Inadequate policies and procedures can result in security failures or censure from regulatory agencies and auditors.

User Authentication
User authentication, correctly identifying the individuals that access the Web site, is another risk area that generates much public discussion. The creation and management of adequate user ID and password files or other authentication measures like digital certificates is a detailed process that requires a large amount of time and attention.

Protection Methods

Considering the risks associated with Web-based financial applications, and taking into account the potential avenues through which risk factors may present themselves, what are the commercially reasonable security methods available today for the protection of Web-based financial application host sites?

User IDs and Passwords
It is not by accident that both the FDIC and the NCSA list data privacy and confidentiality as the first element in their lists of potential security risks. This security component is especially vital to Web-based financial institution activity, since the protection of private personal financial data is one of the cornerstones of the banking industry. Limiting access to information to only those authorized to receive it and transmitting data in a way that ensures privacy are the first steps to successful Internet security.

The most common method of restricting data access on the Internet is much older than the technology that utilizes it. User ID and Password requirements were a key component of earlier technology, such as mainframe computer systems and off-the-shelf PC modem-to-modem communication programs. While the correct application of User IDs and Passwords are still vital to Internet security, their implementation has become more sophisticated by necessity. It is no longer enough to require users to identify themselves—it is now vital to protect against sophisticated programs that can replicate the login efforts of a user, utilizing a pattern of codes designed to uncover the authentic User ID and Password assigned to a specific user. It is vital that Web-based financial applications use measures like limiting the number of incorrect logins and the tracking of unsuccessful access attempts in order to protect the confidentiality of the User IDs and Passwords themselves.

From a procedural standpoint, it is also vital that financial information providers create and implement secure methods for producing and issuing customer codes. User ID and Password controls must become part of the documented, enforced, and audited procedures of every financial institution that maintains more than a marketing presence on the Internet. Separation of duties and dual control within the operations area that maintains user access information and/or the Web site itself must be maintained as ardently as they are in the cash vault or other key areas of the financial institution. The best technology, poorly managed, will prove ineffective.

Data Encryption

Restricting access to sensitive information is the first step toward maintaining data privacy, but financial sites must also take steps to protect this information as it travels across a public, non-secure system. The Internet is by design an open system. It, in and of itself, does not provide secure pathways for private transmission of data. In order for financial institutions to be able to deliver account and transaction summaries, they must utilize encryption methods that will render their transmissions unreadable to parties that are not a part of the conversation. Secure Socket Layer (SSL) and Secure HyperText Transfer Protocol (S-HTTP) are two standardized, widely available technologies to provide encryption of transmissions between Web content providers and browser-based customers. Both of these technologies utilize data scrambling and recompiling methods that render data useless to anyone who does not have the "key" required to read the packet. Most industry-standard browsers support these technologies, and they are relatively easy to implement for commercial Web sites.

The field of data encryption is a constantly evolving area of increasingly complex technologies, some so secure that the United States Government has made it illegal for US companies to export the technology outside of its boundaries. For a more detailed discussion of these methods, see the article "Mmm-mmm Good! Alphabet Soup Security" by Chey Cobb, located in the Library pages of the International Computer Security Association Web site (http://www.icsa.net/).

Certificate Authorities and Digital Certificates

Certificate Authorities and Digital Certificates are emerging to further address the issues of authentication, non-repudiation, data privacy, and cryptographic key management. A Certificate Authority (CA) is a trusted third party that verifies the identity of a party to a transaction. To do this, the CA vouches for the identity of a party by attaching the CA's digital signature to any messages, public keys, etc., which are transmitted. Obviously, the parties involved must trust the CA, and identities must have been proven to the CA beforehand. Digital Certificates are messages that are signed with the CA's private key. They identify the CA, the represented party, and could even include the represented party's public key.

The responsibilities of CAs and their position among emerging technologies continue to develop. They are likely to play an important role in key management by issuing, retaining, or distributing public/private key pairs.

Virtual Private Networks
A virtual private network (VPN) is a private data network that makes use of the public telecommunication infrastructure, maintaining privacy through the use of tunneling protocols and security procedures. Using a VPN involves encrypting data before sending it through the public network and decrypting it at the receiving end. An additional level of security involves encrypting not only the data but also the originating and receiving network addresses.

Routers and Firewalls
On the Internet, a router is a device, or in some cases software, on a computer that determines the next network point to which a packet should be forwarded toward its final destination. The router is connected to at least two networks and decides which way to send each information packet based on its current understanding of the state of the networks it is connected to.

A firewall is the point at which your private company network and a public network such as the Internet connect. A firewall system is a hardware/software configuration that sits between the private and public networks, controlling what traffic is allowed in or out of the private network. Features include logging and reporting, automatic alarms at given thresholds of attack, and a graphical user interface for controlling the firewall.

When properly configured, a network firewall can be very effective at keeping out unauthorized users and stopping unwanted activities on the internal network.

Virus Detection and Prevention
Because potentially malicious programs can be downloaded directly onto a system from the Internet, virus protection measures beyond the traditional boot scanning techniques should be implemented to properly protect servers, systems, and workstations.

Physical Security
The most obvious element of securing a network is often times the most easily overlooked: physical security, or controlling access to the most sensitive components in your computer network, such as a network administration station or server room.

No amount of planning or expensive equipment will keep your network secure if unauthorized personnel have access to central administration consoles. Even if a user does not have evil intent, an untrained user may unknowingly provide unauthorized outside access or override certain protective configurations.

Policies and Procedures
A site security policy is required to establish an enterprise-wide program of how both internal and external users interact with a company's computer network, how the corporate computer architecture topology will be implemented, and where computer assets will be located. The policy should weigh possible threats against the value of personal productivity and also of corporate assets which may need different levels of protection.

Redundancy and Disaster Recovery
Redundancy attempts to eliminate any single point of failure that could disrupt communications on a network. Examples of hardware redundancies include RAID 5 disk systems, dual power supplies in file servers, uninterruptible power supplies, and server clustering/failover technologies.

It is impossible to over emphasize the need for a good backup and disaster recovery strategy. File system backups not only protect you in case of a hardware failure or accidental deletion, but they also protect against unauthorized changes made by an intruder.

The CustomerLink System

CustomerLink is a Web-based application that allows financial institutions to distribute information and accept service requests from their customers anywhere in the world. Since its basic design is geared toward the browser-oriented world of the Internet and Virtual Private Networks, security has been a consideration from its inception.

The CustomerLink system runs on a robust Windows NT operating system. Included in this OS is the capacity to allow only secure connections by end users. Utilizing Secure Socket Layer (SSL) technology, all transmissions of Web pages and data between the financial institution and its customers are completely encrypted and would be unreadable to any person or group that was able to "intercept" the transmission. SSL encryption is industry standard and is commonly used in Internet applications that require security and privacy for sensitive data.

CustomerLink also utilizes user IDs and passwords for all incoming callers. Without the proper login, customers cannot see or use any Web pages within the PC Banking site. Password security is fully under the control of the financial institution, and it allows for implementation of dual control for institution operators.

CustomerLink also includes its own internal security system based on security groups. Groups are security- and fee-level profiles designed by the financial institution. Each customer loaded into the CustomerLink system is attached to a Customer Group that controls what features they may use and what fees they will be charged. Every system operator set up by the financial institution is attached to an Operator Group that provides detailed controls governing the areas of the system that they may view or change. Financial institutions can establish as many groups as are needed to provide the levels of security that they wish to implement.

CustomerLink can be installed as an Internet application, or it can be configured to run on a Private Network. Private Networks, or Intranets, use the same communications technology as the Internet, but access is restricted to only approved individuals. Financial institutions that wish to allow customers to access the system from the Internet or from a Private Network can implement both solutions using the same Web server.

The CustomerLink system is also available in a service bureau environment. Financial institutions can simply provide daily data files to the Equifax-operated data center, and service bureau staff members will handle all aspects of system operation and security. For more information on these configurations, see the diagrams included in this document.

CustomerLink and the Windows NT Environment

CustomerLink was built using the Microsoft BackOffice Suite of server applications to provide a powerful, scalable, and secure Web development platform. The BackOffice products that comprise the CustomerLink Banking System include Windows NT v4.0, SQL Server v6.5, Internet Information Server v4.0 (IIS), Secure Socket Layer v3.0 (SSL), and Certificate Server v1.0.

Windows NT was built from the ground up with the intention of meeting the security criteria for the U.S. Government's C2 Security Evaluation. The critical need for an operating system to be designed for optimum security from the ground up was noted by the NCSC, which wrote in its Final Evaluation Report of the Windows NT operating system: "When security is not an absolute requirement of the initial design, it is virtually impossible through later add-ons to provide the kind of uniform treatment to diverse system resources that Windows NT provides."

The robust security architecture of Windows NT Server is used consistently across all BackOffice components, with authentication tied to controlled access to all system resources. IIS integrates into the Windows NT security model and operating system services such as the file system and directory. Because IIS uses the Windows NT Server user database, administrators do not need to create separate user accounts on every Web server, and intranet users need only to log on to their network once. IIS automatically uses the same file and group permissions as the existing file, print, and application servers.

Some Web servers install their own security implementations on top of the operating system, creating additional overhead and potential security exposure due to lack of integration and synchronization. Windows NT Server is inherently secure by design. Files and system objects can only be accessed with the proper permissions. User and group accounts are managed by a globally unique identification. When accounts are deleted, all access permissions and group memberships are deleted. So even if a new account is created using a previous user name, none of the permissions are inherited.

Certificate Server is a standards-based server application for managing the issuance, revocation, and renewal of digital certificates. Certificate Server generates certificates in standard X.509 format. These certificates can be used for any number of public-key applications including:

- Server and Client authentication under Secure Sockets Layer (SSL) protocol
- Secure e-mail using Secure/Multipurpose Internet Mail Extensions (S/MIME)
- Secure payment using Secure Electronic Transaction (SET)

To run Certificate Server, a Server ID from a third-party vendor is required to register the server on the Internet and allow for SSL security and authentication. Certificate Server will NOT work without a Server ID. Average cost for a ServerID is $350 per year from Verisign.

CustomerLink Security

CustomerLink security is based on the concept of controlled "groups" that allow each financial institution to establish as many levels of security within the CustomerLink system as are required to meet their needs. There are two kinds of security groups: Operator Groups and Customer Groups. Financial institutions can create as many of each kind of group as they desire, creating levels of security that are appropriate for their particular configuration. Financial institutions may also remove options from their scripts and Web pages completely so that no customers can access them regardless of the group to which they are assigned.

Operator Groups
Operator Groups provide a very detailed level of control. For each action within the CustomerLink system, operators can be restricted altogether, be allowed only to view the data, or be allowed to add, edit, and change the data. Two default Operator Groups are delivered with CustomerLink. The Technical Group allows operators to perform all operator actions within the system except for security functions. The CSR Group is designed for operators with very limited system duties related to customer maintenance or retrieval of customer-initiated messages.

Security Officers
In addition to being placed in an Operator Group, an operator can also be designated as a Security Officer. Like operators in the Technical Group, Security Officers are allowed to perform all actions within the system. Security Officers also have the authority to add other operators or change the profile of existing operators. Each financial institution usually has only a single Security Officer, with an additional security login stored under dual control in a secure location.

Customer Groups
Customer Groups provide access control in much the same manner as Operator Groups. When added to the system, each new customer must be assigned to a Customer Group. As customers access the CustomerLink Web pages, they will be limited to performing actions authorized by their Customer Group, and they will be restricted to specific amount limits designated individually for each of those actions. Three default groups are delivered

with CustomerLink. The Main Group allows users to perform all of the actions available. The Restricted Group allows customers to view their account information, but blocks them from performing any transactions within the system. The Demo Group allows financial institutions to set up customers and accounts that can be used by potential system users to see how the system works. Demo customers can perform all transactions within the system, but those transactions are not included in the daily reports and files used by the financial institution. The Demo Group blocks customers from changing the Setup information (such as the Password) related to the demo customer.

Customer Groups also control the amount that users are charged for performing any action in the system. Financial institutions that wish to implement different fee levels for different sets of customers may also utilize multiple Customer Groups for that reason.

Web Page Settings
In addition to Customer Groups, financial institutions can easily remove options from their Web site through the Web Page Settings option in the CustomerLink Administration program. This will make those options, as well as the links and Help files associated with them, unavailable for all customers. This process can be performed by any operator with the proper security rights and requires no special Web page programming or customization.

Glossary of Terms

This list includes terms mentioned in this document as well as a few other concepts and phrases found in the field of Internet security. This list is by no means exhaustive.

Abuse of Privilege:
When a user performs an unacceptable action, according to organizational policy or law.

Application-Level Firewall:
A firewall system in which service is provided by processes that maintain complete TCP connection state and sequencing. Application-level firewalls often re-address traffic so that outgoing traffic appears to have originated from the firewall, rather than the internal host.

All-in-one Box:
A system that contains the set of Internet services required for a fully functional Internet presence, all included in a single appliance.

Authentication:
The process of determining the identity of a user who is attempting to access a system.

Authentication Token:
A portable device used for authenticating a user. Authentication tokens operate by challenge/response, time-based code sequences, or other techniques. This may include paper-based lists of one-time passwords.

Authorization:
The processes of determining what types of activities are permitted. Usually, authorization is in the context of authentication: once you have authenticated a user, they may be authorized different types of access or activity.

Bastion Host:
A system that has been hardened to resist attack, and which is installed on a network in such a way that it is expected to potentially come under attack. Bastion hosts are often components of firewalls, or may be "outside" Web servers or public access systems. Generally, a bastion host is running some form of general-purpose operating system (e.g., UNIX, VMS, WNT, etc.) rather than a ROM-based or firmware operating system.

Challenge/Response:
An authentication technique whereby a server sends an unpredictable challenge to the user, who computes a response using some form of authentication token.

Chroot:
A technique under UNIX whereby a process is permanently restricted to an isolated subset of the file system.

Cryptographic Checksum:
A one-way function applied to a file to produce a unique "fingerprint" of the file for later reference. Checksum systems are a primary means of detecting file system tampering on UNIX.

Data-Driven Attack:
A form of attack in which the attack is encoded in innocuous-seeming data, which is executed by a user or other software to implement an attack. In the case of firewalls, a data-driven attack is a concern since it may get through the firewall in data form and launch an attack against a system behind the firewall.

Defense in Depth:
The security approach whereby each system on the network is secured to the greatest possible degree. This term may be used in conjunction with firewalls.

DNS Spoofing:
Assuming the DNS name of another system by either corrupting the name service cache of a victim system, or by compromising a domain name server for a valid domain.

Dual-Homed Gateway:
A dual-homed gateway is a system that has two or more network interfaces, each of which is connected to a different network. In firewall configurations, a dual-homed gateway usually acts to block or filter some or all of the traffic trying to pass between the networks.

Encrypting Router:
see Tunneling Router and Virtual Network Perimeter.

Firewall:
A system or combination of systems that enforces a boundary between two or more networks.

Host-based Security:
The technique of securing an individual system from attack. Host-based security is operating system and version dependent.

Insider Attack:
An attack originating from inside a protected network.

Intrusion Detection:
Detection of break-ins or break-in attempts either manually or via software expert systems that operate on logs or other information available on the network.

IP Spoofing:
An attack whereby a system attempts to impersonate another system by using its IP network address.

IP Splicing/Hijacking:
An attack whereby an active, established session is intercepted and co-opted by the attacker. IP Splicing attacks may occur after an authentication has been made, permitting the attacker to assume the role of an already authorized user. Primary protections against IP Splicing rely on encryption at the session or network layer.

Least Privilege:
Designing operational aspects of a system to operate with a minimum amount of system privilege. This reduces the authorization level at which various actions are performed and decreases the chance that a process or user with high privileges may be caused to perform unauthorized activity resulting in a security breach.

Logging:
The process of storing information about events that occurred on the firewall or network.

Log Retention:
How long audit logs are retained and maintained.

Log Processing:
How audit logs are processed, searched for key events, or summarized.

Network-Level Firewall:
A firewall in which traffic is examined at the network protocol packet level.

Perimeter-Based Security:
The technique of securing a network by controlling access to all entry and exit points of the network.

Policy:
Organization-level rules governing acceptable use of computing resources, including security practices and operational procedures.

Proxy:
A software agent that acts on behalf of a user. Typical proxies accept a connection from a user, make a decision as to whether or not the user or client IP address is permitted to use the proxy, perhaps does additional authentication, and then completes a connection on behalf of the user to a remote destination.

Screened Host:
A host on a network behind a screening router. The degree to which a screened

host may be accessed depends on the screening rules in the router.

Screened Subnet:
A subnet behind a screening router. The degree to which the subnet may be accessed depends on the screening rules in the router.

Screening Router:
A router configured to permit or deny traffic based on a set of permission rules installed by the administrator.

Session Stealing:
See IP Splicing/Hijacking.

Trojan Horse:
A software entity that appears to do something normal but which, in fact, contains a "trapdoor" or attack program.

Tunneling Router:
A router or system capable of routing traffic by encrypting it and encapsulating it for transmission across an untrusted network, for eventual de-encapsulation and decryption.

Social Engineering:
An attack based on deceiving users or administrators at the target site. Social engineering attacks are typically carried out by telephoning users or operators and pretending to be an authorized user, to attempt to gain illicit access to systems.

Virtual Network Perimeter:
A network that appears to be a single protected network behind firewalls, which actually encompasses encrypted virtual links over untrusted networks.

Virus:
A self-replicating code segment. Viruses may or may not contain attack programs or trapdoors.

Chapter 4: Internet Banking Products

Anytime, Anywhere Banking: Dream or Reality ?

Title: *Anytime, anywhere banking: Dream or Reality?*
Author: *Datanet Corporation Limited*
Abstract: *Anytime, Anywhere Banking is the dream and the goal of the banking industry today. This article explains how it can be achieved.*

Introduction

Anytime, Anywhere Banking is the dream and the goal of the banking industry today. Account holders want it, bankers want it, ..., everybody wants it. Because financial transactions completed faster, without having to physically go to a bank branch, or without being constrained by the so-called "banking hours", would benefit all concerned. In case of personal banking, it would reduce the need to travel to and from banking branches - something especially welcome by customers in our overcrowded and polluted cities. And in case of business customers, the benefit is obvious - it would help strike and clinch deals and help complete commercial transactions faster, and thereby profit all concerned.

Anytime, Anywhere Banking - how ?

How do we convert Anytime, Anywhere Banking from a mere slogan to reality ? Some steps in this direction have been taken in our country.
For example, we hear about Any branch banking having been introduced by one or more banks, i.e., a customer is able to transact his account from any one of a small set of branches of his bank, which has networked these branches. However, the customer still has to go to one of these branches, i.e., he cannot transact from anywhere in the city or country, nor from anywhere in the world. Then there is also the cost of networking, which must be borne by the bank.

Some banks have also introduced the 24-hour ATMs (Automatic Teller Machines). While this makes Anytime a reality Anywhere is still not achieved. Then there is the high cost of the ATMs, and of the computer-communication network, which must be borne by the bank.

Let us now start thinking of how to translate Anytime, Anywhere Banking from a dream to reality. The key words are Anytime and Anywhere. How do we create a communication network, which has a wide reach, and with affordably low cost terminal equipment. Such a network must preferably encompass a whole city, or the whole country, or the whole world.

Furthermore, it must be on 24 hours of the day, and 365 days of the year (366 days in leap years!). The answer is obvious. Such a network already exists; it is the telephone network, or Public Switched Telephone Network (PSTN), as it is called; it spans our cities, our villages, and in fact the whole country. There are about 10 million telephone instruments already installed in India, at a measly price of about Rs.500/- each, and currently about 2 million new telephones per year are being installed. This network has also already been connected by our Department of Telecommunications to the global telephone network, and therefore, a call can be made anytime from anywhere in the world to anywhere in India or vice versa. The network is also on all the time; remember, when the power fails in your home or office, the telephone is still working; this is because every telephone throughout the world is fed by the exchange battery. And the banks do not have to spend one rupee on either the setting up or maintenance of this network or of the telephone instruments.

Now, before using this vast network for banking, or for telebanking, as it is called, there are a few crucial technical issues that must be solved.

Anytime

Before we elaborate on the issues posed by Anytime, let us first consider how telebanking works in the first place. Datanet proposes installation of a piece of equipment called Teleserver in the bank branch or other office of the bank. There are no restrictions on the location of this office, except that one or more telephone lines must be available for connecting the Teleserver.

A customer goes to any telephone instrument anywhere in the world, and calls up the Teleserver to make service requests, such as funds transfer, opening letters of credit, payment of bills, inquiry, etc., The Anytime requirement makes it highly desirable, perhaps even mandatory, that Teleserver should be fully automated, i.e., it should be unmanned. Being fully automated also renders it significantly tamper-proof.

Data input from the telephone instrument

What are the options available for data input? It will be seen that they are (a) the microphone and (b) the keypad, usually with only twelve keys, but with the old rotary dial telephone only the ten digits, 0,1,.....9. Furthermore, when a key on the keypad is pressed, a series of pulses is generated in case of pulse-dialling telephone; or a pair of tones is generated in case of DTMF (dual tone multifrequency) telephone.

Let us first consider usage of the microphone. If one inputs data by speaking into the microphone, then we need to have speech recognition capability at the Teleserver end. Unfortunately, reliable computer speech recognition, especially speaker-independent speech recognition, is still in its infancy and very expensive. This can perhaps be overcome by making the Teleserver a manned equipment, but then we are dependent on the hearing abilities of human operators, which vary from person to person, and also give room for fraud.

Consider next the input of data through the keypad. At the outset, we are handicapped by the availability of only ten keys in general, namely those labelled 0 through 9, giving the impression that only ten different types of characters can be input. (How do we input payee names, bank names, amounts with decimal points in between, etc., with only ten keys ?). However, we can get over this by the method of coding, i.e.., by pressing say two keys to input each character, in which case we are in a position to input as many as 10 X 10 = 100 different types of characters. Pressing any two keys to input each character is a small price to pay in order to get from the simple telephone keypad the capability of a regular computer keyboard!

In Teleserver, data input is done character by character, and through the keypad. Input of each character requires the pressing of two keys. (This does not apply however to the performance of control functions, for which pressing single keys is adequate). The character code is easy-to-remember and easy-to-use. The 26 alphabets, A through Z, are coded 11 through 36, i.e.., you dial 1 followed by 1 in succession for inputting the alphabet A,..., and you dial 3 followed by 6 in succession to input Z. Next the ten numerals 0 through 9 are coded 40 through 49. The code for . (i.e, full stop or point) is 52.

Data output from the telephone instrument

As regards data output, the only option we have available is the receiver, i.e., response from the Teleserver must be voice signals coming to the telephone's receiver. Such signals must be computer-generated, rather than human-generated, so that the Teleserver's response is always uniform, and always the same under similar circumstances. This requirement is met by making the Teleserver a fully-automated piece of equipment.

Security considerations

Authentication

Every customer is assigned a unique identity number, or ID number for short, by the bank branch.

Besides, he is assigned a unique password of adequate number of digits, which he is required to maintain confidential. Additionally, a 4-digit PIN (or Personal Identification Number) is also assigned. The customer is required to remember his PIN and not write it down anywhere. Using the Teleserver, the customer can change his PIN as frequently as he desires, and the same is not accessible to the bank officials.

When the customer dials up the Teleserver, he must input his ID number, his password, and his PIN and if all of these match, he is considered to have authenticated himself, and thereafter, permitted to transact.

A second method of authentication available with Datanet's Teleserver is by use of the Defender series products of Digital Pathways, Inc., This gives us the highest levels of security acceptable in the banking industry worldwide.

Data security (immunity from eavesdropping/masquerading)
Data encryption at the transmitter end and decryption at the receiver end using appropriate keys is now the standard technique used to achieve such immunity. VLSI electronic devices are available implementing the block ciphering DES and RSA algorithms, which are accepted worldwide as providing adequate protection from a malicious intruder for commercial applications.

In the case of the Teleserver, transmission and reception of data at the Teleclient end are character by character, and by a human transactor. In view of these considerations, substitution ciphering is used in Datanet's Teleserver. The Teleclient Management Software running in the Teleserver can assign a distinct private character code for transmit and receive, by the Teleclient, which are unique for each Teleclient. Note that there are potentially 43! or approximately 1052 codes possible, which is a very large number, ensuring a high level of data security.

Malicious access of Teleserver/R data
All data in Teleserver/R is stored in an industry-standard ODBC compliant database. It consists of sensitive data pertaining to Teleclients/customers, such as authentication keys, private transmit and receive codes, and transaction specific information, such as, amounts being transferred, transferee bank and country names, etc., Therefore, the data stored in the database can also be ciphered, again using a substitution cipher. Additionally, access is restricted to a specific bank employee(s) via password and PIN. The PIN can be altered at will by the authorised bank employee(s). The substitution cipher and password are unique to each Teleserver/R shipped out by Datanet.

Conclusion

Datanet Corporation Limited's Teleserver is a near ideal means for achieving Any time, Anywhere Banking. The Teleserver is fully automated, and can accept all kinds of service requests. And it provides the highest levels of security as demanded by the banking industry.

Web-Enabled ATMs

Title: *Web-Enabled ATMs*
Author: *Compaq*
Abstract: *The Internet is transforming the way people live, work, and play. Compaq has developed an innovative solution that enables financial institutions to integrate ATMs as a viable and cost-efficient delivery channel in an unpredictable e-business world.*

Sweeping changes in the financial services industry present both significant challenges and substantial opportunities for financial services providers. Market consolidation, global competition, complex regulatory issues, rising customer expectations, and the emergence of new delivery channels combine to challenge traditional approaches to consumer banking. At the same time, the explosive growth of the Internet and the rise of electronic and mobile commerce have created exceptional opportunities for acquiring banks and processors seeking new sources of business.

While online home banking services and Web-based financial services are enjoying increased acceptance by consumers, automated teller machines (ATMs) continue to provide a cost-effective way to meet customer requirements. In addition, with the institution of ATM usage fees and surcharges for selling products such as postage stamps or entertainment tickets, ATMs have begun to generate added revenues for financial institutions.

Although ATM use among retail customers continues to rise each year, recent studies suggest that less than half of all bank customers use the machines regularly. Shifting additional in-branch business to this cost-efficient delivery channel represents an important cost-savings opportunity for banks. Unfortunately, the explosive growth of mobile and electronic commerce is working against the efforts to move more transactions to ATMs by redirecting transactions to other channels.

While the ATM, as a channel, has achieved its goal of cutting internal bank expenses, another effect—not totally foreseen—is that the branches are losing contact with their customers. This comes at a time when cross-selling, one-to-one marketing, loyalty, co-branding, and customer profitability analysis are of paramount importance.

The key for success in today's rapidly changing environment is to expand the capabilities of ATMs as a self-service delivery channel while making them an integral part of the e-business world. To accomplish this, Compaq—the long-time leader in computing solutions for financial services providers—is offering NonStop™ Net Services, an innovative solution that allows financial institutions to Web-enable their ATMs. With NonStop™ Net Services, banks can easily and effectively consolidate delivery channels—standardizing on a familiar Web browser interface and supported by a fault-tolerant, highly scalable infrastructure.

The growth of electronic self-service banking

At their introduction in the mid-1970s, ATMs represented the financial services industry's first self-service channel. Today, more and more banking is being done through a proliferation of electronic self-service delivery systems.

As the world enters a new millennium, the ATM market still holds considerable growth potential for financial institutions. In 1990, there were only a little more than 325,000 ATMs worldwide. This is expected to increase to approximately 2 million by 2005. With more ATMs available, customer usage continues to accelerate rapidly. While only about 40 percent of customers used ATMs regularly in 1997, that number was expected to expand to approximately 54 percent by 2000.

Part of what is driving this growth is the strong trend toward electronic self-service throughout the financial services industry. To gain or maintain a competitive advantage, banks and savings institutions are constantly searching for ways to reach out to customers where they live, work, and play—24 hours a day, 7 days a week. The rise of automated call centers, PC banking, and Web-based delivery of financial services—in addition to the expansion of ATM networks—all reflect this do-it-yourself movement in the financial industry. In fact, GartnerGroup projects that approximately 30 percent of all customer banking activity today involves electronic self-service transactions.

While traditional ATMs have been both functional and highly cost-effective as self-service cash machines, present- and past-generation ATMs are not equipped to fully capitalize on the substantial growth anticipated in the electronic banking arena. To succeed in the age of e-business, financial institutions need an effective way to increase the capabilities of their ATMs. The Compaq NonStop™ Net Services solution offers an innovative approach.

The effects of e-commerce on the evolution of ATMs

The growth of the Internet and increasingly widespread use of e-commerce is changing the payments landscape, not only for consumers and retailers, but also for financial institutions. The effects of e-commerce on the financial services industry are being felt in a number of areas:

- Creation of individualized products and services. By integrating advanced customer relationship management (CRM) systems with electronic delivery channels, it is now possible for financial institutions to conduct truly one-to-one marketing campaigns—tailoring products and services to meet individual requirements and delivering these

products and services over the customer's channel of choice.
- Control of operations costs. Using the Internet as a de facto communications standard, the Web browser as a common interface, and the Web server as a universal access point, e-commerce serves as a powerful unifying force in driving operational costs down.
- Consolidation of delivery channels and mechanisms. Banks can easily standardize the way products and services are presented to customers. Web pages, home banking screens, interactive (Web) TV, and ATM displays can be created with a common "look and feel," delivered from the same server platform. This allows for important economies of scale, while providing a familiarity that ensures a greater comfort zone for customers.
- Enhanced ATM revenue sources. Financial institutions expect to increase profitability by charging for third-party advertising.

E-commerce has also produced new types of electronic transactions—many of which are card-based—that pose a meaningful threat to the business handled by traditional ATMs. For example, the traditional countertop point-of-sale (POS) terminal at merchant locations is increasingly being replaced by the home PC and wireless devices as more people purchase products and services over the Web. Of course, these electronic purchases are not made with the cash many bank customers withdraw at ATMs.

Stored-value "smart" cards are another emerging form of payment that is generating a growing number of transactions. Some analysts predict that within 10 to 15 years, ATMs will evolve into the channel of choice for individuals who do not own a home PC capable of receiving a secure electronic purse downloaded to a smart card.

Compaq has remained abreast of these trends and is working closely with financial institutions around the globe to prepare products and services that take advantage of the opportunities created by e-commerce. The result is that banks will be able to capture increased revenues, reduce operations costs, and offer enhanced service levels to customers.

Compaq's solution for Web-enabling ATMs

By making it possible to combine efficient, cost-effective electronic delivery systems with the ease and flexibility of a familiar Web browser interface, the Compaq NonStop™ Net Services solution delivers unique advantages.

The Compaq NonStop™ Net Services solution provides the infrastructure for using a Web server and Web browser interface to support financial transactions that originate on ATMs. The solution is a content management and content retrieval application with defined ATM interfaces. Technically current, it is open and runs on Compaq's fault-tolerant NonStop™ Himalaya™ platform. It has been designed to be "put to sleep" in the event of volume surges or time-of-day peaks so that it doesn't interfere with timely transaction processing.

In the future, the solution's combination of a business-critical hardware infrastructure, software, and services could be used to Web-enable any channel a bank desires—from the same Web server. This will not only provide the benefit of giving users access to products and services through a familiar, easy-to-use browser interface, but will also ensure the common look-and-feel desired across a variety of delivery channels.

Compaq's solution is not intended to run on the Internet. Banking transactions and proprietary customer information are routed through a TCP/IP connection with the same reliability, security, and privacy that exist in today's financial networks and within the bank's intranet.

Enabled with Web browser technology, the next generation of ATMs will be capable of a broad range of functions:

- Behaving like full-function POS devices
- Handling new types of transactions such as ticketing and discounts
- Displaying advertising for financial institutions and retail businesses, using dynamic screens, content, and time-of-day management
- Executing one-to-one marketing programs using decision support solutions
- Grouping of messages by common devices, time zones, languages, and so on

By offloading content from the ATMs to a centralized Web server, Compaq's innovative approach greatly improves a financial institution's flexibility and responsiveness to competitive challenges. Web-enabled ATMs can personalize how information is presented to different customers, making each customer feel like he or she is viewed and valued as an individual—recapturing the experience enjoyed by customers in hometown banks in generations past. Further, screens and content to support targeted marketing campaigns and promotions can be developed more quickly and more easily by authorized bank personnel.

Moreover, the NonStop™ Net Services solution enables information from bank customer information systems (CISs) and CRM databases to be downloaded to the Web server for more effective one-to-one marketing to either high-net-worth or at-risk customers. Many financial institutions have had difficulty identifying and communicating with these key customers in the past. Information about these individuals can be stored on the Web server and used to display customized screens to appropriate customers when they use an ATM—translating into improved customer retention and higher revenues and profits.

Finally, because the Web server runs on a NonStop™ Himalaya server, the solution delivers the unsurpassed availability, absolute data integrity, and outstanding scalability and manageability required for a business-critical function such as ATM processing.

Clearly, NonStop™ Net Services can transform ATMs from one-dimensional, cash-dispensing devices into flexible, full-featured delivery channels for a broad spectrum of products and services.

Critical operational and competitive advantages

The NonStop™ Net Services solution for Web-enabling ATMs turns a financial institution's investment in its ATM infrastructure into a fully functional, transaction-rich self-service customer environment. It also provides the unique opportunity of expanding revenue (advertising) and reducing costs (channel consolidation).

The solution greatly expands the spectrum of transaction types ATMs can handle, by converting cash-based transactions into card-based transactions. Customers can visit ATMs to purchase stamps, get tickets to entertainment or sporting events, buy lottery tickets, or obtain similar products and services—generating additional revenue for the bank. Financial institutions, in turn, can essentially act as distributors for certain types of products and services, gaining additional fee income in the process. (ATM interchange fees are income to the acquirer.)

The NonStop™ Net Services solution offers substantial flexibility to financial institutions. The solution is designed to interface with the next generation of touchscreen ATMs running on the Microsoft® Windows NT® Server operating system. This standardization will provide cost efficiencies for MIS groups involved in creating and implementing content and programming. New screens can be developed quickly to respond to even the most fleeting marketplace opportunities, while guaranteeing an appropriate level of security. In addition, acquiring banks can centralize ATM software on their Web server, helping to reduce the cost of upgrading and maintaining ATMs over time.

Finally, as previously noted, the solution allows the same browser-based interface and display screens to be used not only by the Web-enabled ATMs, but also by a variety of other delivery channels, such as PC-based home banking. The consistency and efficiencies that this provides will help banks compete more effectively against an increasing number of technology-savvy competitors.

Key components of the Compaq NonStop™ Net Services solution

The complete NonStop™ Net Services solution includes a NonStop™ Himalaya server with software (including Compaq iTP Secure WebServer software) and a range of professional services to allow for end-user customization. This is why the solution is called NonStop™ Net Services. Financial institutions get not only fault-tolerant hardware and software, but also the technology expertise and applications experience required to deliver total solutions to customers.

The NonStop™ Himalaya server becomes an integral part of the bank's current ATM network, able to use existing financial network interfaces and encryption/transaction security features. Yet the innovative solution won't disrupt performance levels or affect transaction integrity.

At the heart of the solution is a combination of superior content management and ATM grouping logic that allows the right message to be sent to the right ATM at the right time. The solution is deployed as a software developer's toolkit (SDK) to allow for easy customization and to simplify the creation and implementation of advertising and marketing content. This content can be retrieved by using metadata (rules or parameters)—established by the user—to control when and where various messages are sent.

Each SDK release is bundled with "common objects," consisting of a collection of industry-standard tools, application programming interfaces (APIs), and print objects needed to manage specific content or to deliver a particular type of service. The NonStop™ Net Services solution is designed to make it easy to tailor solutions to fit an institution's specific business and technical requirements. With this SDK, financial institutions can more effectively differentiate themselves and gain an edge over competitors in a challenging marketplace.

A rich feature set

The NonStop™ Net Services solution provides a feature-rich technology infrastructure for increasing the functionality of today's ATMs. The solution's key features are supported through five versatile frameworks:

- Content management services. These services provide a secure means of loading content onto the Web server. The content is typically a collection of advertising and marketing messages intended to be sent to the ATM at a specific time. Text and graphics are created using HTML, extensible markup language (XML), Java™ applets, and scripts. The framework also provides a means of updating or deleting content as needed, building a database of content metadata, handling maintenance with utilities for HTML content moves and bulk content moves, and producing reports about usage.
- Content delivery services. This framework determines how content is obtained and delivered to an ATM and tracks content hits and fulfillment requests for billing purposes.
- Customer information services. The NonStop™ Net Services CIS collects usage information for both "on-us" and "not-on-us" cardholders. The ATM is the only existing banking channel that allows a bank to "talk" to customers that are not its own. This unique opportunity has not been exploited for years. NonStop™ Net Services stores the on-us cardholder BIN and ICA prefix table and is able to generate separate messages for bank customers and non-bank customers. Future releases of the NonStop™ Net Services solution will log all ATM transactions so that profiles can be built for both on-us and not-on-us customers, to use in framing the right marketing and advertising messages for each cardholder.
- Financial delivery services. This framework provides host-based financial transaction components and gateways to ATM, POS, and other transaction-authorization systems residing on the host server. Other host-based financial components, including account inquiry, funds transfer, and print statement, can be linked by gateways to all customer accounts at the institution. Because both

the service framework and authorization systems can reside on the same platform, messaging between the two is accomplished via high-performance interprocessing communications that reduce the need for high overhead protocol stacks, compared to other Web-enabled implementations.

- External delivery services. These services provide gateways to other Web or financial services available from the institution, ranging from mortgage and loan services to securities brokerage services. In addition, external or third-party connectivity can be enabled to provide additional goods and services to cardholders. If a financial institution desires, external delivery services can also include Internet access.

Compaq iTP Secure WebServer software is the driving force behind the solution's wide array of capabilities. iTP Secure WebServer software provides high-performance access to information stored in a Compaq NonStop™ SQL database, in addition to application access using Common Gateway Interfaces (CGIs) that are either industry-standard or optimized for a Compaq Pathway environment.

The NonStop™ Net Services solution also deploys iTP Active Transaction Pages (ATP)—an industry-standard, server-based, JavaScript™ environment—for dynamically generating Web pages using NonStop™ SQL database software and Compaq's content management and content retrieval services.

To help MIS personnel make the most of the solution, complete management facilities are coupled with standard utilities for the NonStop™ Himalaya platform. These utilities provide performance statistics and monitoring based on a consolidated view of the environment across systems, and also allow users to make necessary changes to the server configuration. For example, Event Message Services can be used for a variety of operational purposes, including errors and alarms or traps, usage logging, tracing, and business message capture for functions such as billing or fulfillment.

To ensure the security of the Web-enabled ATMs, the NonStop™ Net Services solution incorporates Secure Socket Layer 3 (SSL3) technology to encrypt messaging where required. The process relies on the Compaq WebSafe2 Internet security processor for host-based hardware-assisted encryption, to reduce processor overhead and enhance system response times. This process does not interfere with any hardware-mandated cryptography used to secure financial transactions processed through the payment system.

Finally, the NonStop™ Net Services solution is designed to interoperate seamlessly with both legacy and Internet-style environments, while taking full advantage of the NonStop™ Himalaya platform's traditional enterprise fundamentals of availability, scalability, and data integrity. System solutions can be configured on the NonStop™ Himalaya platform using Compaq NonStop™ Distributed Object Manager, which is fully CORBA 2.2 compliant, with both Java and C++ bindings available. Installation scripts are provided to streamline the implementation process, and full documentation is included. In addition, three-level customer support is available to financial institutions through Compaq's Tandem Business Unit—24 hours a day, 7 days a week.

The initial release of the NonStop™ Net Services solution will include iTP Secure WebServer software and all the SDKs, APIs, and code required to support basic banking transactions, content management, and systems operation. The first release is intended to support up to 100 Diebold ATMs in pilot mode. Future releases will significantly increase the solution's capabilities: Release 2 is intended to support up to 1,000 ATMs, which will include two additional ATM types; release 3 is intended to support up to 15,000 ATMs, also with additional device support.

E-commerce and the Compaq advantage: Doing business with a leader

Compaq is widely recognized as a global leader in providing solutions to the financial services industry. The technology advantages of the Compaq NonStop™ Himalaya platform and the company's unparalleled industry expertise make Compaq the ideal choice for business-critical e-commerce solutions such as Web-enabled ATMs.

Compaq has a proven track record in supplying business-critical solutions to major financial institutions worldwide. The company's Tandem Business Unit provides solutions to more than 700 of the world's leading banks, including 24 of the 25 largest U.S. banks, 18 of the 25 largest European banks, and 18 of the 25 largest banks based in the Pacific Rim.

Today, 80 percent of all worldwide banking transactions and 66 percent of all POS transactions run through Compaq NonStop™ servers. In addition, 18 of 20 U.S. EFT switches and 75 of 100 EFT switches around the world are based on the NonStop™ Himalaya platform.

The reason for the popularity of NonStop™ Himalaya servers is the extraordinary reliability, scalability, and performance that the platform provides.

NonStop™ Himalaya servers are designed to deliver the unsurpassed levels of availability, data integrity, and security required for self-service electronic delivery channels such as Web-enabled ATMs. The NonStop™ Himalaya platform is designed for applications in which any downtime—planned or unplanned—is unacceptable. NonStop™ Himalaya servers stay up and running during routine maintenance and upgrades, support real-time remote backup for protection against environmental disasters, and provide instant and transparent recovery from outages. In addition, solutions running on NonStop™ Himalaya servers protect sensitive customer data that may be downloaded to ATMs, provide absolute assurance of customer identification, and match security procedures to the level of risk involved.

Moreover, as Internet-based and other e-commerce applications such as Web-enabled ATMs continue to expand and transaction volumes soar, NonStop™ Himalaya servers allow financial institutions to economically keep pace with growth. The platform provides powerful performance, plus the ability to expand by an almost unlimited amount—simply by adding processors. Further, linear scalability ensures that banks will receive a full processor's worth of performance for each new processor that is added.

Compaq enjoys "best-of-breed" partner relationships with leading payment system–related software providers, including ACI, eFunds, Mosaic, QSI, Diebold, and others. In addition, Compaq Services—with more than 27,000 experienced professionals worldwide—stands ready to work with financial institutions to develop a Web-enabled ATM solution customized to each institution's unique requirements.

Summary

As customers increasingly look to self-service electronic delivery channels such as Web-enabled ATMs to meet their financial requirements any time of the day or night, financial institutions need a solution that they can rely on without fail: Compaq NonStop™ Net Services.

Customers are becoming more demanding, and a greater percentage of products and services are being delivered electronically. The Compaq NonStop™ Net Services solution enables financial institutions to respond to these challenges by taking greater advantage of the existing ATM delivery channel. It provides for increased revenue and decreased costs, a unique combination.

The solution is highly flexible and can be easily customized to meet the needs of both the financial institution and individual customers or groups. Moreover, it is built on the proven Compaq NonStop™ Himalaya platform, known for its outstanding scalability, reliability, and availability.

Compaq—a global leader in information technology and systems integration—serves hundreds of major banks worldwide, offering "one-stop shopping" for financial solutions. From enterprise servers, desktop PCs, and palmtop devices to world-class software applications, service, and support, Compaq provides a single source for integrated, cross-platform solutions.

For More Information
Website: www.compaq.com

Remote Banking Solutions

Title: *Remote Banking Solutions*
Author: *Compaq*
Abstract: *Remote banking services—often referred to as home banking or PC banking—give financial institutions the ability to rebuild personalized relationships with customers. Compaq offers a series of interesting solutions to help financial institutions realize this.*

Over the past several decades, financial institutions have sought ways to reduce branch operations costs through the use of new service delivery options such as telephone banking and automated teller machines (ATMs). Unfortunately, the byproduct of this effort has been considerably less face-to-face contact with customers—shifting the industry from a relationship focus to a product focus. Remote banking services—often referred to as home banking or PC banking—give financial institutions the ability to rebuild personalized relationships with customers. Retail customers can conduct the majority of their banking business through a variety of electronic channels, including the telephone, personal computers, and remote kiosks over the Internet or proprietary networks. By adopting remote banking solutions, a growing number of banks have reduced the cost of transactions and branch operations, while making banking easier and more personalized for customers. In addition, remote banking is helping financial institutions to retain existing customers, attract new ones, and establish a firm foundation for increased fee-based income in the future.

Compaq's remote banking solutions package proven application software from ACI Worldwide and Edify with enterprise servers and professional services to create turnkey remote banking environments, using a simple and highly engaging interface. In addition, Compaq freely draws on its leadership in business-critical computing to ensure that solutions will perform as expected, without fail—24 hours a day. This allows financial institutions to slash time-to-market for products and services, while virtually eliminating the risks associated with deploying new solutions.

Banking challenges in the 21st century

As the business world prepares for a new century, financial institutions face a number of challenges in retaining customers, increasing revenues, and achieving strategic and operational objectives.

Today's changing financial environment, evolving marketplace, and emerging technologies such as the Internet have given rise to a whole new spectrum of customer expectations and needs. One of the most critical challenges for banks and other financial institutions is to reestablish intimate, personalized relationships with customers—a component of the banking experience that has declined dramatically in recent years.
When most banking was conducted face-to-face at a single hometown office, local bank personnel maintained personal relationships with customers. They knew many customers by name; kept track of the types of accounts, balances, and other services each customer had with the bank; and understood how changing personal or business circumstances might affect a customer's financial requirements. Because branch representatives saw cus-tomers on a regular basis, there was also ample opportunity to suggest additional products or services as needed—from college loans to retirement planning.

Decreasing face-to-face contact
As banks have implemented aggressive branching strategies and turned to technology as a means of increasing service while reducing costs, the nature of financial services delivery has undergone a fundamental change. The frequency of personal interactions between banks and customers has decreased dramatically. In fact, in an alarming number of cases, personal contact has disappeared entirely.

Compaq solutions allow banks to efficiently reestablish personalized relationships with customers while continuing to reduce branch operating costs.

As a result, new accounts representatives and tellers have far fewer opportunities to cross-sell products and services, and often approach sales opportunities with incomplete information about a customer. Credit decisions, too, may be made with little or no firsthand contact and an insufficient understanding of a customer's value to an institution.

Further, in the absence of a personalized relationship, customers have less incentive to remain loyal to their bank and can be more easily swayed by attractive interest rates or prices offered by competitors. This customer attrition impacts a bank's profitability, reduces deposit balances, and costs the bank in additional marketing expenses.

Reduced profitability
The effects of deregulation and increasing competition on spreads have also had a major impact on profitability. To overcome the decrease in interest income, most banks are continuing to work to reduce costs, learn how to market products and services more effectively, and transition to a fee-based prof-itability model. This means creating and promoting value-added services that will eventually generate substantial fee income.

Reestablishing personalized relationships
Compaq's remote banking solutions address the needs of financial institutions in today's changing environment. Compaq solutions allow banks to efficiently

reestablish per-sonalized relationships with customers while continuing to reduce branch operating costs. In addition, remote banking can improve profitability by driving down the cost of transactions.

What's more, as customers become first familiar with, and then attached to, convenient remote banking services, banks will be able to begin charging fees for these services in the same way that they have implemented charges for ATM services. The ultimate irony is that the current wave of technology-based solutions can be used to correct problems that older technology applications created in the past.

What is remote banking?

Remote banking describes a broad range of technology-based solutions that let customers conduct business with a financial institution instantaneously—anytime and from virtually any place. Remote banking solutions describe the next generation of retail banking services. Using personal computers, modems, and other electronic devices, consumers and business people rely on remote banking to conduct a wide variety of traditional banking transactions over the Internet or other communications networks. In addition, some remote banking solutions are accessed via the telephone or by using conveniently located self-service kiosks.

Once a customer connects to a bank's computer network, he or she must generally enter a personal identification number (PIN) and password before conducting any business, which may include

- Traditional banking services—Verification of account balances, funds transfers, bill paying, and credit card and installment loan payments.
- Investment management services—Online trading of stocks, bonds, and mutual funds using electronic brokerage firms such as e.Schwab and E-Trade, delivered through the bank.
- Insurance services—Allowing customers to purchase or make payments for insurance services such as credit liability, personal, homeowners, life, and health.
- Nonfinancial services—Conveniences such as check orders, stamps, and tickets for entertainment events.

Compaq's remote banking solutions

Compaq's remote banking solutions bring together the hardware, software, and professional services that financial institutions need to efficiently and economically deploy remote banking services. Compaq solutions are designed to run on low-cost Web- and Windows NT® Server–based plat-forms and take advantage of the same network infrastructure used for existing retail services such as ATMs.

Compaq's family of remote banking solutions incorporates two emerging technology standards to achieve widespread connectivity and minimize the complexities of dealing with multiple customer devices, network access points, and disparate legacy systems. The first of these is OFX, a collaboration of CheckFree, Intuit, and Microsoft. The OFX standard manages the connection between customers' Web browsers and the bank's server, and defines how data can be exchanged over the Internet. The second technology standard is called the GOLD Messaging Standard. An open, cross-platform interface developed by Integrion Financial Network—a consortium of 18 banks and IBM—the GOLD Messaging Standard facilitates connections to various service providers and makes it easy to add value to existing services.

Compaq's remote banking solutions route customer requests to the appropriate computer system in the bank's data center. The solutions can interface with the channel integration management (CIM) system to consolidate account and customer information from product-centric back-office or legacy systems. They also can interface with customer relationship management (CRM) and decision support systems, to support cross-selling of additional products and services based on customer segments.

Figure 1: Remote banking as an integral part of the retail banking enterprise

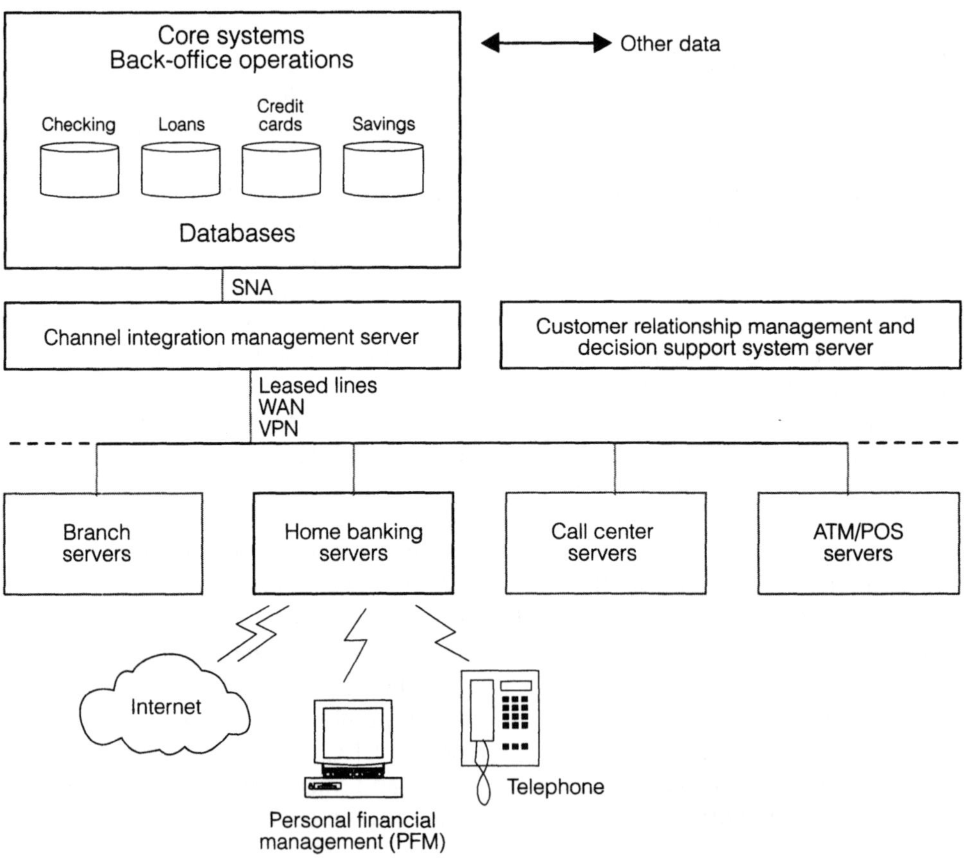

Solution components
The major components of Compaq's remote banking solutions include:

- NonStop® Himalaya® servers—Recognized as the leading platform for business-critical computing for more than two decades, today's NonStop servers deliver the highest levels of availability and cost-efficient scalability for remote banking applications.

- Compaq ProLiant servers and workstations—Compaq offers industry-leading performance and availability for the Windows NT Server platform, including ServerNet® clustering technology to enhance the reliability of the Windows NT Server environment.

- BASE24 software from ACI Worldwide—BASE24-telebanking allows self-service banking and bill paying to be initiated via PCs, the Web, telephones, interactive voice response systems, smart phones, personal digital assistants, and other interactive devices.

- Electronic Banking System (EBS) and Electronic Workforce applications from Edify—These applications simplify the design and implementation of Web bank-ing services via telephone, fax, or PC-based personal financial management software running on the Windows NT Server operating system.

- Atalla Internet Security Processors (ISPs)— Hardware coprocessors provide powerful, hardware-based encryption and decryp-tion of messaging and transactions.

- Compaq professional services—These services can help a financial institution plan, design, and install a solution that

Is the Internet really secure?
For in-branch transactions, banks employ a number of physical security measures, such as armed guards, alarms, and restricted-access areas. Compaq's remote banking solutions employ standards-based digital processes to replicate many of these physical security processes. To prove that customers, merchants, and other financial parties are who they say they are, iTP™ Certificate Security Solution (iTP CSS) from Compaq offers a secure infrastructure that provides full certification authority functions, including management of electronic certificates and keys. Public-key cryptography and digital signatures authenticate and protect senders and recipients, and provide the basis for nonrepudiation of messages and transactions. Signatures and keys are stored using methods that cannot be compromised by an intruder.
To protect cryptographic keys, which are used to unlock encrypted information, Compaq remote banking solutions perform all cryptographic operations in secure, specialized hardware modules from the Atalla Security Products Group. By opting for hardware-based rather than software-based security, orders and purchases can be com-pleted quickly, safely, and privately.

suits its specific requirements. Service offerings include feasibility studies, planning, implementation, migration, consulting, and related support for electronic delivery of all kinds of retail banking services.

Key benefits for banks

Once installed, Compaq remote banking solutions will deliver a number of key benefits to financial institutions:

- Reduce costs—Remote banking transactions cost one-third to one-sixth or less compared to similar branch banking transactions, according to analyst reports. For example, in 1997 the Tower Group found that the average Internet banking transaction costs US$0.08, compared to US$1.40 for a teller-assisted branch transaction.

- Retain and attract customers—Easy-to-use and convenient remote banking services help banks retain customers. They attract more profitable customers because they tend to appeal to technically savvy consumers and business people, who often have a high net worth. These customers expect their banks to offer advanced technology.

- Serve customers better—Remote banking services enhance a bank's service level by reaching customers anytime, anywhere. Combined with an effective decision support system, remote banking services can also help financial institutions create tightly focused marketing campaigns that enable them to cross-sell more effectively and gain wallet share.

- Raise revenues—Many banks plan to eventually charge a fee for remote banking services, following the same business model as ATM services.

Real-world applications

Leading financial institutions worldwide are using Compaq's remote banking solutions to improve their retail service level, control costs, and reach new segments with a variety of electronic products and services.

For example, banks have opened virtual branches on the Internet, taking advantage of low-cost Web technology to offer a full selection of banking services including up-to-the-moment account information, funds transfers, transaction searches, and electronic bill payment. Some banks are also moving into the investment arena, giving customers access to real-time stock quotes, financial planning calculators, merchant malls to serve the needs of small to medium-sized retailers, and electronic mail notices of changes to loan or deposit rates.

Other types of financial services providers are taking advantage of Compaq remote banking solutions as well. A regional network and electronic funds transfer (EFT) processor based on the East Coast of the United States installed a Compaq technology–based solution to successfully allow customers at participating financial institutions to check their balances, transfer funds, pay bills, print statements, or download account information into personal or small-business financial management soft-ware, 24 hours a day.

Count on Compaq

Drawing on the technology at the heart of NonStop servers and an acknowledged leadership in business-critical computing, Compaq's remote banking solutions provide the capabilities needed to compete in this electronic age.

Compaq enterprise servers are a solid choice for the most demanding computing requirements. This makes Compaq solutions ideal to meet the unique demands of remote banking—providing the outstanding performance, unsurpassed availability, and scalability financial institutions will need as their applications grow.

Unmatched industry experience
NonStop servers have an unmatched track record of success in the financial industry. Every day NonStop servers process 70 percent of the world's money (US$1 trillion). NonStop servers currently handle 66 percent of all credit card authorizations and 80 percent of all ATM transactions, and support some of the largest financial institutions worldwide. Major banking industry customers include Bank of America, Capital One, Chase Manhattan, Rabobank Nederland, Royal Bank of Canada, Sakura Bank, U.S. Bancorp, and Wells Fargo Bank.

Industry's best availability
Financial institutions worldwide depend on NonStop servers to provide 24-hour-a-day, 7-day-a-week (24 x 7) availability, enabling financial institutions to offer customers the convenience of banking anywhere, anytime. NonStop Himalaya servers continue running despite hardware failures, software bugs, operator errors, loss of power, or even natural disasters. NonStop servers are engineered to prevent failures and recover rapidly if outages do occur, and they offer a unique, important benefit for remote banking: Should a hardware or software failure occur during transaction processing, when processing resumes, the transaction will continue where it left off instead of beginning again. This feature dramatically improves performance and reliability, and reduces the risk of incorrect financial data. Reducing downtime to near zero also saves money in operations costs. In addition, NonStop servers can undergo reconfiguration without loss or disruption of access to services or information. This means that financial institutions can have absolute confidence in their electronic communications and information systems.

Scalability
NonStop servers also provide the industry's most scalable infrastructure. NonStop Himalaya servers are linearly scalable—that is, a full processor's worth of performance for each processor added. Linear scalability is particularly desirable because the appeal of remote banking is expected to increase dramatically over the next few years. Compaq ProLiant servers, the platform for Edify's Electronic Banking System (EBS), are also highly scalable. In addition, ServerNet technology enables banks to leverage their investment in servers by simply adding new servers, bandwidth, or both, as the number of remote banking transactions increases.

Price/performance
The clustering architecture of Compaq enterprise servers is a cost-effective

combination of efficiency and performance. Clustering divides the workload among many processors, each with its own memory, storage, and system software. As new processors are added to an enterprise sys-tem, their full performance and power are gained without diminishing the performance of existing processors.

Highest levels of service and support
Finally, financial institutions benefit from Compaq's global presence with more than 120 sales and support offices worldwide. By involving Compaq early in the process, Compaq implementation teams can help identify business goals, analyze processes, and pinpoint weaknesses—so a bank can offer customers the service they desire with a solution that fully meets the bank's objectives.

High touch in a high-tech environment
Remote banking services from Compaq can truly supply the high-touch relationship banking financial institutions have been missing, using the latest in high-technology applications. Customers will get the products and services they need, interacting with the bank in an effective, enjoyable manner. At the same time, financial institutions can achieve the cost efficiencies they desire and react to changing business conditions with newfound speed and agility.

MIX
Papier aus verantwortungsvollen Quellen
Paper from responsible sources
FSC® C105338

If you have any concerns about our products,
you can contact us on
ProductSafety@springernature.com

In case Publisher is established outside the EU,
the EU authorized representative is:
Springer Nature Customer Service Center GmbH
Europaplatz 3, 69115 Heidelberg, Germany

Printed by Libri Plureos GmbH
in Hamburg, Germany